THE END OF GENETICS

THE END OF GENETICS

GENETICS

DESIGNING

HUMANITY'S DNA

David B. Goldstein

Yale

UNIVERSITY PRESS

NEW HAVEN & LONDON

Yale University Press books may be purchased in quantity
for educational, business, or promotional use. For information, please e-mail
sales.press@yale.edu (US office) or sales@yaleup.co.uk (UK office).

Designed by Dustin Kilgore.
Set in Yale New and Alternate Gothic types by Integrated Publishing Solutions.
Printed in the United States of America.

ISBN 978-0-300-21939-5 (hardcover : alk. paper)
Library of Congress Control Number: 2021939553
A catalogue record for this book is available from the British Library.
This paper meets the requirements of ANSI/NISO Z39.48-1992
(Permanence of Paper).
10 9 8 7 6 5 4 3 2 1

Cover art: DNA double helix by Erzebet Prikel
http://fineartamerica.com/profiles/erzebet-prikel
www.redbubble.com/people/erzebetth

With apologies to John Legend for repurposing his lyrics,
this book is dedicated to Arav, Maya, Aurora, Thomasina,
Theodora, and Bernard, with love, for you
and all your perfect imperfections.

CONTENTS

ACKNOWLEDGMENTS

It has become reflexive for authors to thank agents and publishers for their support and patience. But its being a reflex does not make it any less necessary or true, at least not in my case. Although writing has long been a private passion of mine, the technical writing my career requires did not teach me much about writing a general-interest science book. To the extent that I have learned something about how to tell a story that is both accurate and accessible, I have my agent of nearly twenty years, Georges Borchardt, to thank. And to the extent that my sometimes discursive storytelling has coalesced into something that looks, reads, and feels like a well-thought-out book, I have the guidance of Jean Thomson Black of Yale University Press to thank. For believing that my books, however many years delayed, will eventually arrive, when even I have had my doubts, I have both Georges and Jean to thank. For providing guidance and catching a few key inaccuracies in my descriptions of human reproductive biology I am grateful to Kate Stanley, and

ACKNOWLEDGMENTS

to Tatyana Russo for help with the figures and bibliography. Three anonymous reviewers caught a great many errors of both fact and tone, highlighted key issues to address, and made the book far more accurate and complete than it would otherwise have been.

THE END OF GENETICS

INTRODUCTION

My career has been focused on identifying the genetic changes that cause disease in individual patients and on trying to use those genetic insights to develop effective treatments that target the precise causes of disease. Over the years this kind of work has been called personalized medicine or genomic medicine, and now increasingly is referred to as precision medicine. Whatever name we've given it, it has been both satisfying and frustrating. Genetics research has progressed more rapidly than many geneticists, myself included, had expected. As you will learn in the pages that follow, from devastating genetic diseases of childhood to later-onset and presumed more genetically complex diseases, such as amyotrophic lateral sclerosis (ALS), kidney disease, and heart failure, the mutations responsible for disease are being identified in ever increasing proportions of affected individuals. Treatments for these genetically resolved diseases, however, are lagging increasingly far behind the genetic discoveries. A recurrent theme has been waves of growing enthusiasm that new, precisely targeted treatments are just around the corner, then followed by crushing setbacks. We are currently in a period of renewed optimism about precision medicine, driven above all by stunning scientific advances in genomics, gene editing, stem cell biology, and pharmacology. But it is important to appreciate that we have been here before.

In the 1990s, many people thought that gene therapy was poised to dramatically improve the lives of a large number of patients suffering from genetic diseases, a lot of which had been newly diagnosed through our growing understanding of the human genome. Numerous trials were initiated in which viruses were used to introduce functional versions of genes that patients lacked because of inherited mutations.

One trial that was destined to change the field was initiated at the University of Pennsylvania and focused on a genetically transmitted disease caused by a deficiency of an enzyme called ornithine transcarbamylase (OTC). The disease is due to mutations in a gene on the X chromosome and affects mainly males, who have only one X chromosome. The mutations impair the function of the enzyme, which helps to break down nitrogen, and people with OTC deficiency can develop toxic levels of excess nitrogen in the body, in the form of ammonia.

In 1998 a young man with OTC deficiency, Jesse Gelsinger, enrolled in the trial and was treated with a modified cold virus. The modified virus was meant to introduce a functional version of the OTC gene to Jesse's liver and therefore restore the normal processing of nitrogen, and thereby eliminate the toxic excess of ammonia. Instead, Jesse's body mounted an unexpected immune reaction to the virus used to introduce the functional OTC gene. Within a day of the infusion of the virus, Jesse showed signs of jaundice, and progressed to multi-organ failure. Within five days he was dead. Later analyses suggested that the virus ended up being distributed much more broadly throughout Jesse's body than expected and contributed to a devastating and ultimately fatal immune reaction. The rapid growth in gene therapy trials came to an immediate end, and it would take more than ten years for public trust and enthusiasm to be regained.

This reminder of the challenges in developing genetically targeted therapies is not intended to imply that the current enthusiasm about precision medicine is unjustified. Indeed, scientific progress in this area has been truly remarkable, and this scientific progress has led to increasingly important advances in treatments targeted to underlying causes of disease. Some of these you will learn about in the pages that follow, and some you will have read about in the news in recent years. But these advances are not coming anywhere close to keeping pace with our exploding knowledge of the underlying genetic causes of disease.

I have been concerned about the implications of this sharp discrepancy for many years. A decade ago, I wrote in *Nature* that the explosive growth in the identification of genes responsible for disease coupled with modest progress in developing treatments for those diseases would inevitably lead to increasing interest in ensuring that the genomes of children do not carry such mutations. Despite my being an enthusiastic advocate of — and, wherever possible, contributor to — precision medicine, the past decade has only added to my concerns about how limited our options remain to effectively treat genetic diseases. Not only do we lack effective treatments for most single-gene disorders; we now know that the challenge is even greater than finding a treatment that can undo the direct effects of the disease-causing mutations. The reason lies in the complexity of human biology.

Let us imagine the discovery of an entirely safe gene therapy

—

3

treatment for a genetic disease. That is, we will envision a therapy that is capable of introducing a replacement for the faulty gene into exactly the cell types that need it. Let us further suppose that the replacement gene is expressed at the right times and in the right amounts and that no immune reaction is engendered. Unfortunately even this achievement, which is still very far away for most diseases, would not be sufficient in many cases. We now know that many human diseases have important developmental components. This means that the presence of the faulty gene early in life alters the developmental trajectory of the child, and even that of the fetus. For such diseases, even a completely effective replacement of the faulty gene post-natally would not represent a complete cure.

For these reasons, while remaining committed to the paradigm of precision medicine, I also remain convinced that many of the most devastating diseases will ultimately be "treated" not by precision medicine but by the careful determination of the genomes of our children. If I am correct in this conviction, in the future parents will be faced with complex and consequential choices about how they wish to adjust the genomes of their children. And to make those choices parents will need to understand the underlying science, so that their decisions can be as informed as possible. What they will need to consider is not only what we know about human disease genetics, which the media are often eager to convey, but also what we do not know. Parents will also need to understand what can practically be done to adjust the genomes of their children and what the consequences of any changes are likely to be. And because it is not an overstatement to say that the decisions that parents ultimately make will influence the future of our species, the purposeful design of the genomes of children is something that should concern us all, whether your reproductive years are before you or behind you.

One conclusion I have drawn from these considerations is that, although human genetics has already advanced far enough to provide parents with information on which they may choose to act, key gaps in our knowledge make it very difficult to anticipate the consequences of the actions likely to become possible. These gaps mean that any reproductive engineering that is performed beyond the most straightforward elimination of strongly acting disease-causing mutations will be performed without a complete understanding of the likely consequences of those changes. This is a prospect that I find deeply troubling, and this book above all represents my best effort to empower non-specialists to develop their own opinions about this most central question for the future of humanity.

In response to this deep uncertainty, I have developed a thought experiment in reproductive genomics to help illustrate the kind of genome engineering that could be entertained in the not too distant future. Throughout the book I will refer back to this thought experiment to help make clear that we *will* have the technological ability of making some kinds of adjustments to the genomes of children, *without* having a matching ability to accurately predict the consequences of those adjustments. You will be in a position to understand this thought experiment more fully later in the book, but as a motivation in the reading that follows, consider the following possibility.

All humans carry genetic variants that are very rare in the general human population. Some of those are brand new in each person, and some are variants that appeared first in their parents or grandparents, or in some fairly recent ancestor. These variants, which natural selection has not yet had the opportunity to examine closely and pass judgment on (as long as they are compatible with life), are often damaging to the function of the protein encoded by the genes they affect. Our simple thought experiment, then, is this: What would happen if a couple were to decide not to transmit any such rare variants? I refer to such an experiment as leading to the "common-variant human," in which every place in the genome where a rare variant might occur is replaced with the variant that is common in the human population — that is, a variant that natural selection has reviewed and judged acceptable.

In preparation for this book, I asked a number of prominent geneticists to assay a prediction about this thought experiment, and found widely varying responses. I do not claim to know the correct answer to the outcome of this thought experiment myself. The main point is to convince you that we will be able to actually perform this experiment in the future, and that we will be able to do so long before we know with any confidence what the outcome will be.

The title of this book is of course modeled on the title of Francis Fukuyama's book *The End of History and the Last Man,* published in 1992, an elaboration of his essay "The End of History" from 1989. Fukuyama argued that liberal democracy was a sociocultural end-state, and with its ascendancy, the evolution of human social organization would come to an end. The experiences of the present, with an apparent march toward authoritarianism in many parts of the world, would suggest that this thesis is demonstrably wrong, as most political scientists would today likely agree. Yet there were compelling reasons that Fukuyama made his argument. We can perhaps hope he will in some way turn out to be right.

My title is meant in the same way. I do not mean that human genetics will end. But there are reasons we can argue that the form human genetics has taken until now will shortly come to an end. Here are some clear examples. Human genetics is today and always has been a lottery. We all carry mutations that can cause recessive diseases. For each one, there is a 50 percent chance that it will be transmitted to a child. If it is combined with another such mutation, the child will have a recessive disease. Beyond these mutations that are found in a child's parents, each child born will have its own collection of brand-new mutations, which will appear in the sperm or egg that led to the child. For example, if we consider only "point mutations" that change one particular site in the genome from one thing to another (these "things" changing will soon be fully explained), each human born today has between fifty and a few hundred brand-new mutations distributed throughout his or her genome, with the exact number depending very strongly on the age of the father. If the child is lucky, these new, or de novo, mutations, will fall into parts of the genome that have little or no function and will do no serious damage to the child's development and functioning through life. Each embryo conceived, therefore, is playing an unknown and deadly serious game of genomic Russian roulette, with parents implicitly hoping that all these new mutations will fall somewhere harmless in the genomes of their children. It is a game of chance that all couples through all of time have been obligated to play in order to have children. This is how we have reproduced and how we still reproduce today. It is not how we will reproduce in some tomorrow. And that tomorrow will come.

I believe that it is now certain that parents will soon, and increasingly, decide on the genetic makeup of their children in ways both large and small. When this begins on a large and systematic scale, it is going to constitute a whole new kind of genetics. In consequence, human genetics as it has been since our beginnings will be over. Of course, a new kind of genetics will replace it. What exactly this new kind of genetics will be is for us all to decide. And I believe that everyone should be in a position to participate in those decisions. I hope that this book will help.

Chapter 1

THE FUTURE OF REPRODUCTION

The first decade of the twenty-first century marked a turning point in the relationship between society and the human genome. For the first time it became possible to determine the genetic makeup of any person in a matter of days and at a cost already within range for many millions of people. Even before this genomic watershed was reached, a movement had emerged to provide genetic information directly to consumers.[1] In some cases the offerings to consider included help to make "more perfect babies."[2] The obvious question upon reading such a claim is this: What would qualify as a more perfect baby? Until recently, this question was of primarily academic interest. Ethicists could debate the benefits and the risks, but there was no realistic prospect of systematically prohibiting the appearance of unwanted mutations in our children. For better or worse, the genomic engineering of future generations has suddenly become a very real prospect and is therefore something, I believe, that society must urgently consider. There are two key questions about such reproductive genomic engineering that need to be considered and understood: What do we want to change in the genomes of our children, and what are we likely to be able to change?

In terms of what we would like to change, in my experience, most people are comfortable with the idea of ensuring that children do not carry mutations that cause devastating childhood diseases. Indeed, there are already ongoing efforts throughout the world to reduce the transmission of severe childhood genetic diseases such as Tay-Sachs disease. Many couples already opt for a procedure called carrier screening, which seeks to identify genes for which both mom and dad carry mutations that, when combined, would result in a severe childhood genetic disease. And many fertility centers offer the option of testing embryos for mutations that would cause such diseases in order to select for transfer those that do not carry two such mutations, among other genetic testing options.

But how far should these efforts go? Is it also appropriate to test for mutations that predispose one to, but do not deterministically cause, later-onset conditions such as Alzheimer's disease? And what about non-disease traits, such as height or eye color or, indeed, traits such as sexual orientation, where the genetics is often only probabilistic but where those probabilities can in principle be tested for and parental preferences acted upon? Even if we restrict attention to disease, we face the question, What is or should be classified as a disease? These questions become particularly acute when we recognize that different people will have very different ideas about what they would like to see in the genomes of their children. And views about what is and is not a disease have changed in important ways over time. As one stark illustration, a certain minority of the deaf community has a preference for deaf children. What if a couple wants to use genomic technologies to ensure that their children would also carry the same deafness mutations that they themselves carry? In the past, professional geneticists could draw a certain comfort from technological constraints. As our ability to identify the mutations that make us different from one another grows along with our ability to not only select but, as we shall see, edit what is present in the genomes of our children, what we should and should not do will emerge as one of the defining questions for societies and individuals.

The second part of the question about what we might change in the genomes of our children relates to what kinds of traits we are able to influence and how much we can influence them. When the personal genomics movement was first gaining traction, it was heavily focused on common genetic variants, mainly for technical reasons related to how genetic studies at the time were performed. These common variants are conventionally considered sites in the genome where there is a common form and a minor form, and the minor form is observed about 5 percent or more of the time. This represents a tiny fraction of the variable sites in the human genome. As you will learn in some detail through the course of this book, there are around three billion different positions in the human genome, and we know that most of these three billion sites vary in one or more humans alive today. As you will also learn in the chapters that follow, while we know that most of these sites will vary somewhere in the human population, we still do not know the consequences of the vast majority of that genetic variation.

In the early days of personal genomics, there were two clear drivers of consumer interest: genetic ancestry and the results of "genome-wide association studies," or GWAS. Genetic ancestry is just what it sounds like — your genetics telling you something about where your ancestors are from in a geographic sense and, sometimes, something about relatives you may not have known you have.

And GWAS is a fancy, and somewhat overstated way of describing a type of genetic experiment that allows the assessment of whether any of the common variants in the human genome influence a particular disease or trait. A typical GWAS might, for example, compare a million common variants between patients with schizophrenia and those without schizophrenia to see if any of them are associated with an increased risk of disease. As of this writing, many thousands of different common variants have been associated with hundreds of different diseases and traits, including, by now, almost all the common diseases and many other traits such as height and weight, skin and hair color, and even complex behavioral traits such as educational achievement, IQ, and sexual orientation. All of this can be reviewed using a catalog of GWAS findings.

In the case of ancestry testing, consumers could learn something, albeit rather coarse, about the geographic origins of their Y chromosomes or their mitochondrial genomes, reflecting, respectively, their paternal or maternal ancestry. Or they might get a composite picture of a kind of average of the various geographic ancestries represented in their entire genomes, learning, for example that their ancestry overall appears to be from some part of Europe, Asia, Africa, or the New World. Or they could learn that they are related, to some degree, to someone else who has also been tested. In the case of comparing their genomes against the results of GWAS studies of diseases and other traits, consumers might learn that they have a marginally greater or lesser risk of type 2 diabetes than the population average or that they carry stronger risk factors for certain autoimmune or neurodegenerative diseases. In most cases, however, there has been no reputable advice that could be offered as a function of the genetic information gleaned from such "genomic profiling." For example, one of the very strongest effects for any common variant in the human genome is due to variation at the ApoE gene. Those who carry the risk forms of this gene have a greatly increased likelihood of developing late-onset Alzheimer's disease, but there is currently nothing that carriers can do to meaningfully alter their risks of developing the disease.

This lack of clinical impact led me, in a *New York Times* interview in 2008, to sum up the entire enterprise as "recreational genomics."[3] Referring to the field as recreational was not only intended to convey that what was being discussed in the guise of "personal genomics" was of little to no clinical value but also reflected a degree of exasperation that I (and many other geneticists) feel about how many personal genomics companies implicitly or explicitly oversell what they offer and, in consequence, badly mislead the public about the nature of human genetics. Evidence of how personal genomics companies oversell and misinform is not hard to find. Sometimes it is subtle; sometimes it is egregious.

Alistair Moffat is a journalist and former rector at St. Andrews College in Scotland. He is also the former chief executive officer of Britain's DNA, a now-defunct for-profit personal genomics company that once offered consumers a variety of DNA tests. In the summer of 2012, Mr. Moffat described some of what Britain's DNA had discovered. He first claimed that a volcano seventy thousand years ago "blew itself to smithereens" and destroyed all the human genetic lineages except for those of two individuals, dubbed Adam and Eve. He went on to claim that Britain's DNA discovered a remarkable individual who has Eve's DNA, that they had found a genetic marker from "Queen Sheba," that 33 percent of the men of Britain carry the "founding lineages of Britain," and that 97 percent of men with the surname Cohen share a genetic marker. The earlier description of what some personal genomics companies offered as "recreational genomics" can hardly cover excesses like this. An English geneticist, Mark Thomas, who was involved with me in some of the work that provides the grain of science behind these absurdities, more accurately described them as genetic astrology. Astoundingly, Mr. Moffat's initial reaction to being challenged about the accuracy of these claims was to threaten legal action against Dr. Thomas and one of his colleagues, Dr. David Balding, a highly respected and talented mathematical geneticist. Mr. Moffat's science-free musings would almost be amusing if their potential consequences were not so serious.[4]

The egregious misrepresentation of genetics for what seems to be commercial gain reflects an increasingly complicated relationship between professional geneticists and a public eager to know what the vaunted genomics research might mean for them in terms of their personal histories and the diseases they might face. When the personal genomics craze got started, many professional geneticists felt there was little that was really useful for anyone to know from having their genomes profiled in the ways possible during the dawn of personal genomics. It was very awkward, however, for geneticists to insist that people have no reason to learn anything about their genetic makeups if they are interested in doing so. Insisting that people should not be interested, when they clearly *are,* reflects a degree of paternalism that many geneticists would be reluctant to embrace.

The transition toward sequencing human genomes on a mass scale, however, moves us suddenly into an entirely new dynamic. Much sooner than most geneticists were expecting—certainly sooner than I expected—we suddenly had the capacity to identify mutations of major effect wherever they are in the genome and in whomever they occur. Already, this capacity is filtering through to the personal

—

genomics companies, which are increasingly offering information about mutations that could really make a difference to people. For example, 23 and Me, a generally much more sober personal genomics company than Britain's DNA, got itself into hot water several years ago by offering consumers information that has explicit medical relevance, in particular in testing for a panoply of disease-relevant mutations that might, for example, increase the risk of breast cancer or that might cause an inherited genetic disease in children if coupled with another mutation in the same gene. The FDA ordered 23 and Me to stop marketing their test because it offered medical advice without ever having shown that the test worked as advertised, leading to a lengthy negotiation between 23 and Me and the FDA about the right way to convey such information. We are clearly now moving into an era in which getting things wrong really matters. Although I support the right of people to learn about their genomes if they wish to, I have little confidence that personal genomics companies, largely unregulated at present, can be generally relied on to provide accurate information. It should be clarified that the most significant concerns currently apply to mutations that have a very strong impact on disease, not ones with more subtle effects. These kinds of major-effect mutations do, however, appear to be responsible for a surprisingly broad range of genetic diseases, though certainly not all diseases. Two examples make the new reality clear.

A number of years ago I began a research program at Duke University using a then brand-new sequencing technology to try to find the causes of childhood genetic diseases that could not be resolved any other way. Bertrand Might was one of the first children we studied. Bertrand suffered from an unusual genetic condition that has many similarities to known disorders involving how sugars are added onto proteins but with a striking distinctive feature: When Bertrand cried, he did not make any tears. Bertrand had what we call an undiagnosed genetic disease. My colleague Dr. Vandana Shashi, an experienced and talented clinical geneticist, was sure Bertrand had a genetic disease, and equally sure that it was new. Nothing quite like it had been described before. The Might family had spent the first years of Bertrand's life shuttling from doctor to doctor so that Bertrand could undergo genetic and other tests to find the underlying cause of his condition, but with no success from this wrenching diagnostic odyssey. In 2010 Vandana invited the Mights to join a research study in which we planned to sequence the genomes of twelve patients like Bertrand and their parents and to find *all* the mutations present in the portion of each patient's genome that encodes proteins—the most important part of the genome in terms of disease-causing mutations.

Six months after the Mights entered the study, we had the answer. Matthew, Bertrand's father, carries a very rare and destructive mutation in the N-glycanase gene, which encodes a protein that strips sugars off other proteins. But because Matt also has a normal copy of that gene, he is fine. Bertrand's mother, Cristina, carries a different extremely rare destructive mutation in the same gene. But she also carries one normal copy of the gene and is, similarly, fine. When we sequenced this family, we found the mutations in Cristina and Matthew and discovered that, tragically, Bertrand got both of those destructive and rare mutations. He does not make any functional version of this protein at all. When we performed this analysis the responsible gene, NGLY1, was not yet known as a disease-causing gene. But we felt confident enough to diagnose Bertrand as the first patient with a previously unknown disease related to how defective proteins are removed from cells. We arrived at that conclu-

—

sion, in part, because analysis of other human genomes suggested to us that people should not be found without any working copies of the protein this important gene encodes.

This kind of approach, of learning what parts of the genome do and do not carry certain kinds of functional genetic variation, my colleague Andrew Allen of Duke and I eventually developed into a new formal tool for genome interpretation that we referred to as "intolerance scoring" (which I will describe in chapter 4). The point for now is that we had looked through the entirety of Bertrand's protein-coding genome, and those of his parents, and were able to identify the two ultra-rare mutations that are responsible for Bertrand's condition. Because our inference was based on the evaluation of only a single patient's genome, however, we explained to his parents that we could not be certain of our conclusion until other patients similar to Bertrand were identified with similar mutations in the NGLY1 gene. Then we would know. We reassured Bertrand's parents that we would look hard for such confirmation cases, and the parents reassured us that they, too, would help to find confirmation cases. Of course, we nodded indulgently at this suggestion and assumed that nothing would come of the parents' efforts. But today, we know better. In what has become one of the more dramatic illustrations of how important it is that families be treated as real partners in the study of rare diseases, it was above all the efforts of the family and not those of us leading the research study that quickly confirmed that the diagnosis was correct, as accurately described in a 2014 article in the *New Yorker* on the Might family. Since that time, Matthew Might has become one of the leading global figures in precision medicine.

Tragically, just as this book was being completed, Bertrand "Buddy" Thomas Might died on October 23, 2020, at the age of twelve. But not before he had, without any hyperbole, changed the world, in ways large and small. On the global stage, Bertrand helped to change legislation facilitating patients' access to novel treatments, and his story inspired international efforts in precision medicine. And for the many of us lucky enough to know Bertrand and his exceptional parents, his story has affirmed and strengthened our commitment to do whatever we can to find ways to treat the most serious genetic diseases. The way I now approach my own work traces back in no small part to the hours I spent sitting at a café on the marina at Herzliya Pituach north of Tel Aviv, poring over our first genetic analyses and eventually convincing myself that the NGLY1 mutations in Bertrand's genome must be responsible for his condition. I remember clearly how it felt to have an answer for a child who had gone so long without one, and also the decision to communicate our conjecture to his parents. I felt then, as I do now, that there can be no higher calling for a geneticist. To those of us who have been part of Bertrand's story, it was moving to learn that his father had shared with him the day before he died important news about progress toward a treatment for NGLY1 deficiency. As Bertrand's father related, he was able to "tearfully yet joyfully" relate news he had received from another NGLY1 parent about important progress in the search for a therapy (described online at bertrand.might.net). It is a comfort to think that Bertrand knew how much he had personally done to make real this hope of so many.

The second example is, in one critical way, even more striking. This study began when an eighteen-month-old girl presented at Duke University Children's Hospital with a devastating and progressive neurological condition. No one knew what she had, but they knew that it was very serious, and there was concern she would not have long to live. Lacking a clear diagnosis, the clinical team, again led by Vandana, made two decisions, one of which would shortly change everything for the patient and her family. First, they decided to initiate treatment based on the possibility that the patient could have had an autoimmune condition. As they were unsure of that diagnosis, however, the second thing the clinical team decided to do was to ask our group to sequence the patient's genome to try to determine a genetic diagnosis. Twenty days after receiving blood samples from the patient and her parents, we found that she did not have an autoimmune disease after all. Instead, her genome showed us, with absolute clarity, that she has a very rare genetic disease called Brown-Vialetto-Van Laere syndrome, which results from a broken transporter of vitamin B2. Critically, the way this rare disease is treated is by dietary supplementation with vitamin B2.

The fundamental reason this treatment works is that biology is always messy. The job of the transporter is to move vitamin B2 into the interiors of cells. Without the transporter, not enough of the vitamin gets into cells, and that is what causes the disease. But if there is enough of the vitamin circulating, some gets into the cells even without the dedicated transport system. Only two days after this diagnosis, vitamin B2 supplementation was initiated for the patient, leading to an almost immediate lessening of her symptoms. Not long after the initiation of this treatment, arrived at entirely thanks to modern genomics, Cara Greene, a toddler at the time, ran around an examination room high-fiving members of the genetics team who had performed the analysis. Her remarkable progress was documented by ABC News in 2016 in a segment that has provided inspiration to many affected by genetic diseases. Amid our own astonishment and joy, we could not help wondering how many other patients there are like this who could be helped so very much by

—

this new science. As you will see in the chapters of this book, the answer to this question is both encouraging and discouraging.

As I finish writing this book in spring of 2021, we know of more than four thousand genes that cause one or more genetically "simple" but often devastating genetic diseases. The encouraging news is that children who suffer from one of these diseases are now, in the richer parts of the world at least, increasingly evaluated with cutting-edge genetic analyses to find the underlying causes of their diseases. This still leaves a lot of genetic diseases undiagnosed. A surprisingly high number of adults with what are assumed to be more complex diseases in fact have simpler genetic diseases that are not diagnosed because adults are more rarely evaluated genetically. In poorer parts of the world, even children with apparent genetic diseases often go undiagnosed when what is now an inexpensive and fast genetic analysis could readily provide a diagnosis, raising an important challenge to ensure that precision medicine is made more inclusive than it is today. Even in wealthier parts of the world, medical practice has still not fully caught up with the value of diagnostic sequencing, which is a sobering illustration of the fact that it can take time for practice to change sufficiently to make effective use of new technologies.

I was recently struck by a personal story that made clear how much work remains to be done to ensure that those who need diagnostic sequencing have access to it. Dr. Liz Cirulli is a very talented research geneticist and a former PhD student of mine at Duke University. A year ago, she and her husband took their daughter, Ivy, for an evaluation because of excessive salt cravings. Following the identification of urine abnormalities, Dr. Cirulli's literature review suggested that her daughter might have either Bartter or Gitelman syndrome, each a rare kidney disorder. Her doctors, however, were skeptical, given the rarity of the conditions and inconclusive laboratory results, and did not pursue genetic testing for a possible diagnosis of either syndrome.

As it happens, however, Dr. Cirulli and I taught a course titled the Past and Future of the Human Genome in the Department of Biological Sciences at Duke, together with co-instructor Dr. Misha Angrist. In order to personalize genomic information for the students, I sequenced all three of our genomes in the Center for Human Genome Variation at Duke, allowing us to discuss the particulars of our genetic makeups in the last lecture of the class. After considering her daughter's presentation, Dr. Cirulli reviewed her own genetic data and found that she carries a loss-of-function mutation in the SLC12A3 gene. SLC12A3 encodes an ion transporter and, relevant here, is a known cause of Gitelman syndrome. Given that Dr. Cirulli herself carried an extreme mutation in exactly the gene responsible for Gitelman, her daughter's doctor pressed the insurance company to test at least this one gene. After a considerable period of denying the claim, the insurance company eventually agreed to a test for the SLC12A3 gene alone. Strikingly, the diagnostic lab reported not only Dr. Cirulli's mutation but a second rare mutation that Ivy could have inherited only from her father and had previously been reported as present in another patient. These results suggest a genetic diagnosis of Gitelman syndrome.

It is shocking to me that in 2020 it can be so difficult to get the health system in the United States to provide the genetic evaluations necessary to support the effective diagnosis and management of rare diseases. And given that this is true even for a family member of a prominent geneticist who has already had her own genome sequenced, it is not hard to imagine how challenging this is for many families in similar situations that are not already in the world of genetics.

Addressing both of these substantial limitations in the applications of clinical genomics is an urgent priority for our field. We also know that for an important proportion of the time that we believe a genetic answer should be found, nothing is apparent, raising a critical question for the field: Where is the missing genetic answer to be found? Nevertheless, for many types of diseases, genetic analysis can find a clear single-gene cause in as many as half of the individuals evaluated.

For those lucky enough to be studied genetically and to have the responsible genetic changes identified, however, an obvious question is this: How many genetic diagnoses would permit the development of a treatment as effective as what is available for Brown-Vialetto-Van Laere syndrome? The treatment for this disease, administration of lots of vitamin B2, is a hallmark example of what we now call precision medicine. Although different scientists and clinicians have varied definitions of precision medicine, my definition is simple: Precision medicine involves basing effective treatments on the exact underlying mechanistic cause of disease in the individual patient being treated. Following this simple definition, the vast majority of patients diagnosed with rare genetic diseases today cannot be effectively treated in the way that patients with Brown-Vialetto-Van Laere syndrome can be treated.

Despite remarkable advances in treatment modalities, including even treatments targeted right to the responsible gene, effective therapies are currently available for only a tiny proportion of known genetic diseases. And if a targeted treatment is not already available, we know from experience that the development of targeted thera-

—

pies can take a very long time, sometimes running into decades. And even more challenging, as I emphasized in the introduction, we know that many of the diseases are what we call developmental, meaning that treating them after the onset of symptoms is unlikely to result in a complete cure. I believe that this reality means that many of these known genetic diseases will not end up being effectively treated by precision medicine therapies but will instead be prevented from appearing in the human population by the genomic selection and engineering of the genomes of children. And while the specific diseases are generally rare, severe genetic diseases are, collectively, surprisingly common. Also, many of these diseases that will afflict children are knowable in advance, before a couple even conceives.

Indeed, most people carry several mutations that could combine with other such mutations to cause severe genetic diseases. We call diseases like these, which have very clear familial patterns and are due to very high-impact mutations in specific genes, Mendelian diseases. They are named for Gregor Mendel, who first discovered the basic rules of genetic inheritance that these diseases follow. How Mendel came to invent the science of genetics you will come to understand, in detail, in the next chapter. And while genetics can seem inaccessible, due in part to sloppy popular descriptions, its core principles are in fact not only fairly straightforward but also beautiful. I think you will agree once you have been properly introduced.

Well-known examples of Mendelian diseases include Tay-Sachs and cystic fibrosis. In fact, the number of such mutations that we each carry has been estimated, considering only recessive diseases (those for which two defective copies are required to cause disease). Callum Bell and colleagues, writing in *Science Translational Medicine* in 2011, studied 437 genes that are known to cause diseases that are severe and present in childhood if a child carries two mutations.

The researchers sequenced these genes entirely in 104 healthy people. They found that, on average, each person carries 2.8 disease-causing mutations in these 437 genes alone. If, for simplicity, we assume that mating is random (that is, partners choose each other independent of their underlying genetics) and that any one of these genes is equally likely to carry a mutation in a given person, then for these 437 genes, 2 percent of couples would have disruptions in the same gene. Because the mutations are recessive and two copies are needed to cause disease, a fourth of the children from such couples would be affected. And in order, I hope, to provide a motivation for this entire book, the science behind these numbers that I report to you here without justification will be fully explained in the chapters that follow. After finishing this book you will be able to understand exactly the basis on which scientists make such assertions about genetics, and also be able to judge those assertions yourself, based on the underlying scientific principles.

For now I emphasize that all of these diseases, caused by mutations in those 437 genes, and indeed in many other genes that carry mutations causing recessive and severe childhood diseases, can be prevented. They can be prevented completely, using only currently available technologies. How? Simply by sequencing the entire genomes of prospective parents and determining if they both carry mutations in any of the same genes, considering these 437 genes causing recessive diseases or any others (there are over two thousand genes in total currently known to cause recessive genetic diseases). For those couples who carry mutations in the same gene, a number of practices already tested could prevent the births of children affected by these diseases. Prospective parents could, for example, choose not to have children with a partner carrying a disease-causing mutation in the same gene, as already happens in many religious Jewish communities where marriages are arranged, in some cases informed by genetic information, specifically to prevent Tay-Sachs disease. In other cases, partners may choose to have children together despite knowing they carry mutations in the same gene but use either pre-implantation or post-conception genetic testing procedures to ensure that they bring to term only babies that have been confirmed not to carry two defective copies of the same gene.

Screening for such mutations in people and prospective parents is now called "carrier screening" and is increasingly common in wealthier parts of the world. Today this often happens later than it should, when pregnant mothers discuss the option with their care providers. Before long, it will be performed earlier, allowing prospective reproductive partners to evaluate genomic compatibility by sharing their genomes and checking for genes with mutations in common. Without doubt, in time there will be smartphone apps that permit such compatibility checks while couples await their drinks at a bar. The majority of such tests we know will come up showing the couples compatible, and a minority will flag genes where they both carry mutations. What the couples do with the information will of course rightly be their choice, but they will know in advance. This is on the way.

It is very clear to me that now that we can sequence entire genomes, such screening procedures, and many others, will rapidly become routine, at least in those demographic groups or those countries that can afford it. These procedures will indeed rapidly reduce the burden of inherited genetic diseases in the human population. This is not to say that this first step of eliminating inherited recessive diseases will be easy to implement, will be implemented without errors, or will be fairly accessible to all. But we know how to do this and it is clear that, over time, this sort of testing will be practiced with increasingly greater accuracy and broader application. It is, I fear, equally clear that public interest in such screening efforts will not be restricted to these clear-cut cases of mutations that cause childhood disease. The research community is very rapidly developing knowledge of inherited risk factors that considerably change the risk of not only Mendelian diseases but also more common and complex ones, such as autism, schizophrenia, cardiovascular disease, epilepsy, intellectual impairment, and many others. Will some prospective parents, having had their genomes sequenced by one "personal genomics" company or another, decide that they do not wish to transmit a specific constellation of genetic variants they carry that have been associated with increased risk of developing epilepsy or heart failure or some other condition? Surely they will.

And who is it that will tell them what would be reasonable to test for, and possibly change, and what would not be reasonable? If history is any guide, once the technical ability to engage in such efforts is achieved, there will be many who choose to use it. From my perspective, the only reasonable course of action is to ensure that "consumers" are empowered with as much information as possible to make fully informed choices. We are, in the 2020s, already past the transition from recreational genomics and are entering into a kind of personal genomics that will have very real consequences for future generations. Part of these very real consequences of the genomics revolution will surely involve more advances in precision medicine like the one that so benefited Cara Greene, and that we all rightly celebrate. But my view that progress in human genetics will ultimately lead to widespread engineering of the genomes of children is not a uniformly popular perspective, for a number of reasons. Most fundamentally, this means that much of human genetics will eventually relate not to guidance about how to treat genetic disease but rather to guidance about what not to transmit. Human genetics will transition into a discipline focused on "search and destroy," which is not nearly as appealing a narrative as finding the cause of a nasty disease and thereby curing that disease. And this perspective leads, inevitably, to a paradigm that could accurately be described as eugenic.

Eugenics, for many good reasons, is very scary because of history. The excesses and pseudo-science that characterized twentieth-century eugenics have rightly horrified contemporary geneticists and made many of them deeply hostile to anything that has the feel of eugenics. The term itself conjures historical connotations of Aryan supremacy, along with selective sterilization and its horrors. And it intersects with what I consider to be scientifically crackpot theories of the genetic superiority of some racial or ethnic groups over others, some promulgated very recently.

For example, only ten years ago, two anthropologists, Greg Cochran and Henry Harpending, argued, among other theories, that Ashkenazi Jews have undergone processes of natural selection that have made them genetically more intelligent than other ethnic groups. That different ethnic groups perform differently, on average, on standardized cognitive tests and on other measures of intellectual achievement, such as who shows up in Stockholm to receive that most coveted of scientific prizes, is a matter of fact. Of the more than nine hundred people who have won a Nobel Prize, around 20 percent have been of partial or full Jewish ancestry. This compares with a population representation that is at least an order of magnitude lower in the United States and the United Kingdom, the two countries with the largest share of Nobel Prizes. Cochran and Harpending argued that this discrepancy can be explained by the idea that natural selection has made Jews smarter than other groups. Professional geneticists today, nearly without exception, see absolutely no evidence supporting this position.

When the *New York Times* journalist Nicholas Wade wrote a book provocatively titled *A Troublesome Inheritance* supporting related conjecture, one hundred professional geneticists (myself included) crafted a response emphasizing the complete lack of scientific evidence for such arguments. To those geneticists, the explanation of these factual differences are clear. What you do and what you achieve, including a visit to Stockholm for a Nobel Prize, depends, no doubt, in part on your genetics. But achievement also depends critically on how you were raised. And different ethnic groups, for all sorts of reasons, are raised differently from one another, on average. And they are also, on average, treated differently by others around them beyond their families, an issue that is receiving increased attention today in the context of concerns about systemic racism. Yes, a lot of Jews go to Stockholm. But Jews are also raised differently from non-Jews. And children in poorer parts of Ethiopia and Nigeria and Brazil and Paraguay are raised differently on average from children who grow up in New York and attend schools spectacularly good at science education. The Bronx High School of Science alone boasts eight graduates who have won a Nobel Prize. But no one has yet argued that this exceptional school has changed the genomes of its alumni, as impressive as those individuals are, as I can personally attest, knowing some of them. It is a good school, and that matters. As does the number of books in the house you grow up in, among other factors. And these environmental contributions to a person's development are quite sufficient to explain differences among ethnic groups in measures of cognitive excellence. Such environmental determinants of achievement are well documented, whereas the genetic claims of the superiority of one group or another are explicitly conjecture only.

To make this point plainly, let me state clearly here, as a professional geneticist, that there is not one scintilla of reliable evidence I have ever come across that suggests that differences in the average cognitive performances among racial and ethnic groups have genetic bases. Put more simply, as I was quoted as saying following the publication of Cochran and Harpending's claims, in explaining the unusual intellectual achievement of Jews, I would assign responsibility to Jewish mothers, not Jewish genes.

And yet, despite this sometimes horrific history of eugenics, in a narrow technical sense that is where we are headed. And if we are indeed headed there, it is better to be careful and thoughtful about it as opposed to being squeamishly unwilling to consider the topic because of historical abominations. Admittedly, this perspective is my own, and it is a perspective that some professional geneticists would surely contest (some strongly so). It is, however, my belief that the current generation of young people will be one of the last to reproduce without explicit and widespread consideration of the precise genomes of their children. And this act, of considering and eventually sculpting the genomes of future generations, could be accurately called eugenics. Indeed the simple act of prohibiting disease-causing mutations from being transmitted to children is eugenics. Eugenics is, after all, following a dictionary definition, simply trying to ensure the most desirable characteristics in the genomes of children. It is about selecting what we believe are better genetic forms. And in some cases, this choice is clearly both noncontroversial and desirable, such as not having a child with Tay-Sachs disease, in which severely affected children will live only into early childhood.

—

Of course, we must also recognize that such efforts to engineer the genomes of children are sure to extend beyond the elimination of Tay-Sachs and similar diseases. For these reasons, I would like to persuade people to recognize what is inevitably coming and to convince them of the need to learn enough about human genetics to help them become responsible stewards of the human genome.

But historical connotations matter. Although I maintain that the topic must be considered carefully and broadly, the reality is that the word "eugenics" itself is a loaded term that carries with it a history of pseudo-science often deployed to advance explicitly racist agendas or to promote the separation of the human population into the genomically good and bad. And as much as I believe in the potential of the science this book will explore, the term "eugenics" for me is personally painful because of historical connotations and even more recent usages in support of agendas replete with bias and animus. I believe this is true also for many readers. For this reason, while I have clarified for the sake of accuracy that much of what I discuss could be called eugenics, I will not use the term further in this book. I choose to reject the word not because of technical inaccuracy, but because of the connotations it carries and because of the history of pseudo-science that surrounds it. Instead, throughout this book I refer to efforts to ensure that the genomes of children are as free of disease-causing mutations as possible as *reproductive genome design*. And I will emphasize that in attempting to provide non-scientists with a real understanding of human genetics, I hope, among other things, to arm people with the information they need to recognize racist and biased agendas masquerading as science. I should also admit that the term "design" connotes more competence than we are likely to own in the foreseeable future, as will be made clear in the pages that follow. Therefore, I propose this terminology as descriptive of the aims of the likely changes that will be made in the genomes of children, even if our efforts to design the genomes of the future end up surprising us, as I feel sure they will.

One caveat is in order, related to the question of bias, before moving on to the science. People are today, sadly, increasingly accustomed to politicians and others telling highly biased, factually dubious stories for personal gain. Scientists do have somewhat more constraints imposed on what we claim by processes such as peer review, meaning that comical falsehoods are perhaps more rare among scientists than, for example, politicians. But the narrative that many scientists treasure and promote, that they focus on truth above all, is to me quite clearly a self-serving falsehood. It is clear that careers in science are much like any other careers in many ways. And scientists will, of course, also angle for advantage and often press their own points of view for varied and complex reasons, notably including self-interest. Given this reality, it would probably be fair to say that the biases that influence the claims of scientists are, at a minimum, more subtle than those of some other professionals, but hardly absent for being more subtle. It would therefore be wise to presume that scientists will often, consciously or unconsciously, present evidence and arguments that are designed to buttress and support positions they have staked out in the past. After all, if the positions associated with a scientist carry the day, that scientist will be in a better professional position than otherwise. In this context, it is important for readers to know that for some of the controversial topics covered in this book, I can and should be considered partisan.

I have at times taken strong positions in debates about the nature of the genetic variation that is most important in influencing human disease. In particular, I am aligned with a view that the variations that are most important to many human diseases are relatively rare in the human population and generally "harmful," as opposed to harmful in a more nuanced fashion. This may seem at first pass a rather arcane distinction without much real-world import. In fact, it is central to the thesis of this book, as I will explain in detail in chapters 3 and 4. If the mutations that cause and predispose to the worst of the genetic diseases are usually sharply deleterious and often recognizable as damaging to some known function encoded in the human genome, identifying them and targeting them for non-transmission to children is both a desirable and now an increasingly achievable goal. These kinds of mutations are generally pretty rare, precisely because they are uniformly either very or at least somewhat harmful, and therefore selected against. On the other hand, if the mutations that predispose one to most of the diseases people will suffer are nuanced and subtle in their effects and are actually beneficial in some contexts and bad in others, identifying them and targeting them for non-transmission is both harder to do and of less certain benefit if done. So in describing the science that supports the positions taken in this book, there will be a number of places where some and perhaps many colleagues would not consider me objective. I should therefore be up front and emphasize that some of my perspectives may be influenced by the scientific camp with which I am aligned.

Being in this way associated with a scientific "camp," what can I say to reassure readers? Only this: While I cannot reasonably claim objectivity, I do propose to follow, as far as I am able, a kind of science journalism that includes a differentiation between what are generally accepted scientific conclusions and what are personal judgments and opinions. Insofar as possible, my aim will be to describe the scientific evidence objectively. And where I am aware that I am venturing into personal perspectives, these will be clearly labeled as such, to make it evident that they should be treated as "op-eds." I am committing to you to do my best to keep these separate. But I alert readers to my "party affiliation" to help you judge my assertions for yourself. A related admission is that I have opted, as will already be apparent, to often refer to my own work where relevant. I have made this choice because I believe it is more transparent to let the reader know that material discussed relates to my own work and for the practical reason that I know my own work better than that of others. Because this can sometimes read as self-aggrandizing, even to my eyes, I should make clear at this early stage, without undue modesty, that my own view of my role in human genetics is of one who has generally been a competent practitioner, but not one of the few who has truly changed or led the field. My work appears in many places because I have been involved in many aspects of human genetics for many years, and I like to think my work has been solid and generally (though certainly not always) correct. But in the vast majority of cases my work has contributed to the field, as opposed to changing or leading it. My contributions have varied in importance, but almost always have been within frameworks that are well established. So my views should be seen as, at most, coming from the trenches of human genetics research, not from its more rarefied command centers.

Chapter 2

LEARNING TO READ THE HUMAN GENOME

A warning for geneticists is in order for this chapter. Elsewhere in this book, I have tried to make the material accessible for non-specialists while still providing some novelty for professionals, if nothing else at least in the form of personal views with which they may well disagree. This chapter is somewhat of a departure from that balance in hewing reasonably closely in content to what could readily be found in any number of popular histories of genetics. I have opted to include this material anyway, despite its somewhat more generic content, because I have tried to distill the particular material that is most relevant to understanding the arguments to come. This introduction is therefore mostly an invitation for the in-formed reader to skim over this chapter if so inclined. The only small counterargument I might provide is that I found, once I reviewed early genetics research, that I knew the details much less well than I thought. That might, of course, just be me.

—

The Invention of Genetics

The man who first discovered the rules of genetic inheritance was born in 1822 in what was then the Austrian Empire and was named Johann Mendel. After studying philosophy and dealing with repeated illness and financial challenges, he became an Augustinian friar, adopting the name that history knows him by, Gregor Mendel, and entering the St. Thomas Abbey in Brno, in what is now the Czech Republic. As should be reassuring to anyone who feels they have more to offer the world than may be reflected by standardized tests, Mendel then failed the examinations to become a high school teacher and returned in 1853 to the abbey, first as a teacher and eventually as abbot. During his studies in natural philosophy, Mendel was exposed to research in plant and animal heredity, and in 1856 he began his own work in an experimental garden at the abbey. In just eight years, Mendel used this garden to invent, out of whole cloth, as it were, the modern science of genetics. By either good fortune or good judgment, Mendel concentrated his attention primarily on the edible pea, though he dabbled with several other organisms, including bees and mice; the latter work was deemed inappropriate by the abbey because of the involvement of sex between animals. Little is known about how these explorations led him to his focus on peas.

However he got there, the pea was a particularly good choice for studying the inheritance of observable characteristics, in large part because Mendel could precisely control parentage. He did this by transferring pollen with a brush either to the same plant (leading to self-fertilization) or to a different plant (leading to cross-fertilization). Before beginning his investigations, in a decisive step allowing his discoveries, Mendel carefully observed the variation in his pea plants. Through self-fertilization and assessment of the characteristics of the resulting plants, Mendel identified a number of characteristics that were consistent within individual plants and consistently different among them. He eventually settled on seven traits to study, including seed shape and tint, pod shape and color, flower color and location, and height. Mendel then set out to discover the rules governing the inheritance of these traits.

At the time of Mendel's work, the prevailing view of inheritance was blending inheritance, in which the character observed in offspring is the average of that observed in the parents. This view was derived from observations of many traits that appear to follow such a pattern—for example, human height—where a child's height can be quite accurately predicted as a sex-adjusted average of the height of his or her parents. This prediction is in fact extremely accurate as long as both parents and offspring are well fed, and indeed the violation of this expectation is one of the most striking illustrations of the importance of environment in development. In the decades following World War II in America, it would not have been uncommon to see an immigrant couple of short stature walking alongside their American child towering over them, the European accent of the parents and American accent of the child quickly explaining the discrepancy. The couple would have been exposed, when young, to the deprivations, or sometimes tortures, of the war years in Europe, with their genetic potential stunted in starkly observable form due to malnutrition. The child, in contrast, raised in wealthy America, reflected the genetic potential of the family. But for parents and offspring in the same environments, blending the inheritance of height does very much appear to be the rule. As we shall see later in the chapter, an appearance such as this can be very deceptive in terms of the underlying mechanism.

—

Mendel himself quickly determined that the characteristics he studied in peas dramatically refuted the expectations of blending inheritance. For example, when Mendel performed controlled crosses between plants with wrinkled seeds and plants with smooth seeds, he did not find seeds of intermediate smoothness. Instead, he found that a cross between smooth and wrinkled seeds always led to smooth seeds looking just like those of the smooth-seeded parent, not blended at all. And most of the crosses he performed showed the same pattern — the progeny looked like one parent or the other for the characteristics considered, not ever like an average between them. In terminology that all geneticists still use to this day, Mendel referred to his parental plants (with different characteristics) as P1 and P2 and the first-generation cross between them as the F1 generation. And what he observed in general, as with seed shape, is that the F1 looked like either P1 or P2. And this observation led Mendel to his first law, the *law of dominance and uniformity*. Here Mendel postulated that when the parental strains differ, the F1 hybrid generation will be uniformly identical to one of the parental strains, and he considered the characteristic in the parental form that was observed in the F1 hybrid plants the dominant trait.

From here, the obvious question is this: What happened to the determinants of the "hidden" characteristics? In the all-smooth-seeded peas of the F1 cross between smooth and wrinkled peas, was the determinant of wrinkled peas lurking somehow unexpressed in the hybrid plants? The way to assess this was, of course, for Mendel to return to his ability to cross plants of his choice. So at this point, to address the question of what happened to the hidden trait, he self-fertilized the F1 cross of the wrinkled and smooth parents. And what did he find? Most remarkably from the vantage point of the middle of the nineteenth century, the progeny of self-fertilized F1 plants,

—

37

denoted by Mendel as the F2 generation, resurrected both the smooth and wrinkled forms of P1 and P2. Moreover, and emphasizing the importance of quantitative description of experimental results, these plants were observed in a ratio of approximately 3:1 smooth to wrinkled.

Clearly these observations are flatly inconsistent with blending inheritance, as Mendel well recognized. Blending inheritance should not cause the F1 generation to look like only one parent, and it certainly should not lead to recovering exact replications of the grandparent strains in any of the F2 plants, let alone in a fixed ratio as he observed for smooth and wrinkled seeds and other traits he studied as well.

Mendel needed a new model of inheritance. And it is here that the modern science of genetics was invented. For whatever combination of reasons, though invented here, it did not register in the scientific community for decades to come. Mendel postulated two key components in this new model. One was the elementen, as he called it: a heritable underlying substance of some kind that determined the trait under study. Two, the forms that the elementen could come in, which he called particles. In modern terminology, we know the terms of this model that Mendel conjured to explain his data as genes and alleles, respectively. Working, however, a century before Watson and Crick's elucidation of the structure and function of deoxyribonucleic acid (DNA), Mendel knew nothing whatsoever of DNA, not to mention genes and alleles. But he had in fact discovered genes and alleles, and he had given them names. Again, inventing notation we use today, Mendel referred to his particles with upper- and lowercase letters reflecting whether the characters the particles encoded were the ones that would dominate or would hide in the F1 generation. Explicitly, uppercase "A" was the form that was observed, or was dominant, and lowercase "a" the form that hid, or was recessive. Thus Mendel's parental strains with smooth seeds were AA, his parental strains with wrinkled seeds were aa, and his F1 hybrid strains were Aa, with the A particle dominating the effects of the a particle at the "seed-shape" elementen or gene.

Introductory courses in mathematics always emphasize the importance of clear labels for all variables. Without clear labels, you cannot derive results from rules. Mendel knew enough of mathematics to understand this basic requirement of a mathematical analysis, and his naming of elementens and particles allowed him to address the outstanding quantitative question in his data: Why did the plants segregate in a ratio of roughly 3:1? To explain this pattern, Mendel postulated the *law of segregation,* providing that the particles of the seed-shape elementen pass through to the sex cells (or gametes, pollen and pistil in plants, sperm and egg in humans) independently and randomly. From this postulate, you can quickly derive an expectation of a ratio of 3:1 smooth to wrinkled in the F2 generation, noting that smooth is dominant. To see how this works, return to Mendel's notation with A the dominant trait (smooth, which dominates the appearance of the F1 generation), and the recessive trait (wrinkled, which is hidden in the F1). If the F1 generation is a combination of the two, and if the law of segregation holds, half of the pistils (female plant gametes) are A and half are a, and half of the pollen (male plant gametes) is A and half a. If these combine randomly, four outcomes occur with equal probability— AA, Aa, aA, and aa. Since A is dominant over a, the first three plants present with the dominant smooth appearance, while only the fourth group, aa, is wrinkled (figure 2.1). And since these are expectations only, which can deviate in any particular set of plants, you have to have looked at enough plants to see that the pattern really is 3:1.

—

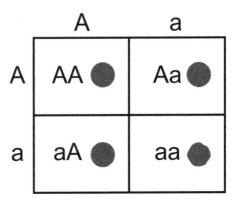

Figure 2.1. Simple hybrid cross. (Drawn by David B. Goldstein and Tatyana Russo)

In the few years that Mendel spent creating the modern science of genetics, he considered one final question, again using the nomenclature he invented and the seven pea characteristics he studied, leading to his third and final law. What happens if you cross plants that differ in more than one of the seven characteristics? For example, consider the cross between plants that are smooth and green with ones that are wrinkled and yellow (green and yellow seed pods was another of the seven traits, with green dominant over yellow). We know already what members of the F1 generation look like. They reflect both dominant traits—that is, all-smooth seeds with green pods.

But what happens now in the F2 generation? What pistil and pollen gametes are produced? If we assume that not only the particles (or alleles) segregate randomly (50/50) at each elementen but that the particles at the two elementens (shape and color) assort independently, we can derive the expected proportions of the multicharacter F2 generation. For clarity, we now switch to R for round, r for wrinkled, G for green, and g for yellow. The pollen will produce the forms RG, Rg, rG, and rg, while the pistil will produce the same set of forms. And these will all combine with one another to produce sixteen different combinations in terms of the underlying particles. If this occurs randomly and independently (what is happening at the shape elementen does not influence what is happening at the color elementen), we can derive the expected proportion of plants in the F2 generation in each possible appearance group just by counting up the combinations that lead to each appearance. Whenever there is an R particle, the seeds will be round, and whenever there is a G particle, the pods will be green. When there are only r particles, the seeds will be wrinkled, and when there are only g particles, the pods will be yellow.

If you count up the relevant groups in the resulting combinations of particles (figure 2.2), you will find that nine of the sixteen boxes show the dominant traits for both elementens (round and green), three out of sixteen show the dominant trait for shape and not color, three out of sixteen show the dominant traits for color and not shape, while one out of sixteen shows the recessive form for both traits. From this expectation and the observations of his plants, Mendel derived his final law, the *law of independent assortment*. The particles at the shape and color elementens were being passed on to pistils and pollen independently of one another.

Exactly how Mendel arrived at these laws we do not know. He must have derived them in large part by developing a theory that fit what he was observing in his plant crosses. But there must have been a good deal of inspiration in coming up with an underlying theory to explain the observations, and we have at least some reason to believe that Mendel emphasized inspiration in part over observation. We can say this because an eminent statistical geneticist, Ronald Fisher, whom we will meet later in this chapter and in more

—

41

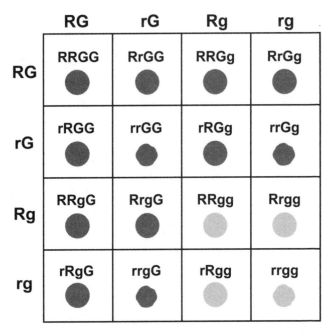

Figure 2.2. Dihybrid cross. (Drawn by David B. Goldstein and Tatyana Russo)

detail in the next chapter, argued that Mendel's results were literally too good to be true. Writing in 1936, Fisher argued that the reported results in peas accorded to Mendel's three laws matched expectation even more closely than would be expected by chance.[1] In making this claim, Fisher deployed a novel statistical approach that has become standard in experimental science (meta-analysis of results across multiple experiments). It might sound contradictory to say that results of an experiment can match expectation too closely. Consideration of simple thought experiments, however, quickly reveals the conceptual argument.

Imagine that a friend of yours were to claim that every one of the ten coins in her pocket is a fair coin—that is, it has an equal probability of landing on its head or its tail. You might challenge her to prove it by asking her to flip each coin ten times. For each of these ten experiments, the expected outcome is five heads and five tails. Now let us assume that she proceeds to flip the coins in turn. The first coin is tossed ten times and lands heads exactly five of those times and tails the other five times. As does the second coin, which again lands five times heads and five times tails. This happens all the way through the tenth coin. Does this outcome prove your friend's coins are fair coins? Hardly. It proves, rather, that your friend has manipulated the tosses. While five heads and five tails is what is expected for each set of ten tosses per coin, the odds of obtaining those results *exactly* each and every time are very low. In formal terms, we say that the probability of obtaining such a result by chance is less than one in a million, meaning that you could ask every person in Manhattan to perform the experiment and you would not expect more than a single such outcome. This, in simple form, was Fisher's charge against Mendel.

Whether this really means that Mendel doctored his results to fit his theory, we still do not really know, and modern scholarship has debated Fisher's argument at length. Depending on exactly how Mendel performed his experiments, the results could be too good to be true, or he might have followed more complicated statistical designs than assumed by Fisher that would not require doctoring for explanation. Whether Fisher's charge is accurate in some way or not is perhaps of some historical interest, but cannot in any way detract from Mendel's discovery. By some unknown combination of careful observation and inspiration, Mendel discerned the laws of genetic inheritance with what is in retrospect breathtaking insight and accuracy. Perhaps early observations of his data led him to postulate an accurate theory that he subsequently validated with some

—

preferential measurements. But, however he arrived at his laws, those laws do indeed govern the natural world, and they govern the underlying causes of disease that now afflict hundreds of millions of people, as we will learn in subsequent chapters. For example, the most common mutation that causes cystic fibrosis is referred to as deltaF508 and occurs in three to five out of every hundred people of north European ancestry. If a man and a woman both carrying the deltaF508 mutation were to have children, on average, a quarter of their children would have only normal copies of the gene, half would have one normal and one mutant copy, and a quarter of their children would have only mutant forms of the gene. Since the deltaF508 mutation is recessive, a quarter of the couple's children will have cystic fibrosis. Sound familiar? It should. The inheritance of cystic fibrosis follows Mendel's laws precisely.

In fact, all four thousand genes that we now know about that cause severe inherited disease follow these laws exactly (with some minor variations, as we will note below). And this, of course, is the reason why geneticists call these diseases Mendelian diseases. They are Mendelian because the pattern of inheritance of the disease itself in families follows precisely the patterns of inheritance that Mendel described in his peas. What we currently call complex human diseases do not obviously follow this pattern. If you look at who has a heart attack in large families, or who has asthma, depression, or some kinds of cancer, you rarely find any patterns that clearly reflect Mendelian inheritance. But if you look at cystic fibrosis, sickle cell anemia, or Tay-Sachs, or thousands of other such diseases, the inheritance patterns are observably Mendelian.

Of course, we now know of many exceptions to Mendel's laws when examining how organisms look. Some alleles (Mendel's particles) do not show consistent expression across individuals; some combinations of alleles at a single gene do not look exactly like either parent but like a combination, something we call co-dominance. Furthermore, some genes (Mendel's elementens) are located near one another, and so the alleles at those genes do not in fact assort independently of one another, while some genes are located on the sex chromosomes and thus show a distinct pattern of inheritance. But in terms of the basic principles of inheritance, Mendel had it exactly right, and we know that at the genetic level, Mendel described the rules that govern health and sickness in the human population.

We also know now, thanks to the quantitative work of Ronald Fisher, that the inheritance rules postulated by Mendel are not incompatible with the apparent blending inheritance of traits like human height. Fisher showed that if there are enough independent Mendelian genes at work, the distribution will end up creating the appearance of blending between the parental trait values. Thus, if a large number of "taller" and "shorter" genes in the genomes of relatives were combined, the result would look like a blending of the heights of those relatives.

Remarkably, although Mendel's work was published in 1866, it had little impact on the scientific community until the beginning of the twentieth century, when his rules were finally rediscovered and popularized. Various reasons for this spectacular lag have been suggested, but none are particularly compelling. Whatever the reasons, it was not for lack of need for the theory. In fact, the most effective contemporary argument against Darwin's theory of evolution by natural selection was provided by Fleeming Jenkin, who demonstrated that blending inheritance would rapidly eliminate variation in the population, thus removing the material that Darwin's accurate theory of natural selection was supposed to act upon. The rediscovery of Mendel's rules in the early twentieth century was therefore one of the essential steps in the eventual development of our modern understanding of both genetic inheritance and evolution by natural selection.

The First Genetic Map

My aim in this book is to provide a sufficient grounding for you to understand what can and will be done in reproductive genome design. Given this narrow focus, I am required to be highly selective in what is covered, making the historical progression seem much easier and sequential than it really was. For those interested in the details of how scientists worked out the rules governing the flow of genetic information and molecular biology more generally, the best accessible treatment remains the masterful (and masterfully titled!) popular science work *The Eighth Day of Creation,* by Horace Freeland Judson. *The Eighth Day* tells the detailed story of what is now considered the golden age of molecular biology. For our more narrow purpose, what we need to understand in addition to the rules of Mendelian inheritance are the basics of how genetic information is encoded, how it is transmitted between generations, and how the genetic information we all carry is translated into the proteins that build organisms. I now turn to the key developments responsible for the modern understanding of genetic inheritance and its translation into organisms. While I am generally following the chronology of these developments for convenience, the organizing theme is how genetic information is encoded, used to build organisms, and transmitted across generations.

Following Mendel's work, it would be thirty years before three European botanists working independently of one another in experiments with plant hybrids rediscovered the rules Mendel had developed. All three, having made their discoveries, uncovered Mendel's earlier work and presented their own as a confirmation and rediscovery of the rules that Mendel had correctly inferred decades earlier. Following the rediscovery and dissemination of Mendel's rules, scientists turned their attention to the identification of the physical substance that carried the instructions first identified by Mendel, which he referred to as elementens. The first question that needed to be answered was where they were to be found.

Embryologists, working with sea urchins, grasshoppers, and other species, determined that body cells carry paired chromosomes with all but one pair matching. They also found, however, that sex cells carry half as many chromosomes and only unpaired ones. Moreover, they showed that normal development requires a full complement of chromosomes. These observations led to the chromosome theory, positing that Mendel's "genes" resided on chromosomes. Experimental proof came a few years later in the famous "Fly Room" of Columbia University. While Columbia takes justifiable pride in its early association with the Fly Room, part of the reasoning behind selecting flies is less a matter of pride. Starting his work at Columbia University, then as now physically constrained by its presence in New York City, Thomas Hunt Morgan wanted a study organism suitable for his cramped laboratories. He rapidly settled on the fruit fly, which was to become one of the most important of all experimental organisms in genetics. Given its short life cycle, ease of breeding, and small size, it was apparent to Morgan that it would allow careful experimental manipulation at scale. While Mendel had worked with tens of thousands of peas in his career, Morgan and his team would work with many millions of fruit flies.

The breakthrough observation came early in Morgan's Fly Room and involved a naturally occurring mutation changing eye color from the normal red to white. Following Mendel and crossing white and red lines, Morgan first observed the white eyes in males only. Following a further series of crosses Morgan eventually deduced that the determinant of eye color must occur alongside the determinant of sex. Coupling this observation with the earlier observations of a chromosome that correlated with sex, Morgan reversed his own recent skepticism and fully embraced the chromosome theory. In so doing, he introduced the first sex-linked genetic characteristic and the first definitive linkage of a genetic trait with a chromosome. And it occurred to him that if some genes are associated with chromosomes, then presumably all would be.

The second bombshell for folks in the Fly Room was that even for visible mutations that were not associated with sex, Mendel's math did not work. And many of the traits did not follow Mendel's law of independent assortment. Instead, some of the different visible mutations tended to occur together more often than Mendel's law asserted. As often happens in science, the breakthrough realization came to one of the youngest members of the group, Alfred Sturtevant, who at the time was a graduate student. It was Sturtevant who realized that the tendency of mutations to be found together in progeny might relate to their physical separation on chromosomes. Noting experimental evidence that paired chromosomes interact in the precursors of sex cells, he reasoned that mutations on the same chromosome might be moved between parental chromosomes based on their physical distance. If this were true, the mutations that occurred most often together would be the closest to one another on a chromosome. By counting such co-occurrences for the many different mutations under study in the Fly Room, Sturtevant built the first genetic maps, inferring the linear ordering of genes on chromosomes. This relationship between the frequency with which mutations on the same chromosome mix between parental chromosomes (cross over) and the physical location of genes on chromosomes is in fact the basis of how the vast majority of genes responsible for serious human diseases were identified, as we shall see in chapter 4.

Human Genetic Literacy: Cracking the Code of Life

With the demonstration that genes are arranged linearly on chromosomes, the next critical step was to identify the physical substance of heredity. The work to identify what genes are actually made of began with an entirely different research effort by the British physician and bacteriologist Frederick Griffith. In an effort to develop a vaccine against pneumonia caused by *Streptococcus pneumoniae,* Griffith worked with two recognized forms of the bacterium. When grown in a petri dish, some of the bacteria formed clusters of bacteria with rough edges, leading to the designation R for rough, while other bacterial strains formed rounded and smooth clusters of bacteria, and were designated S for smooth. When mice were infected with the R form no symptoms developed, while when infected with the S form the mice developed pneumonia. Thus, the R form was determined to be benign and the S form virulent for unknown reasons. Continuing his efforts to find a vaccine, Griffith killed both the R and S forms with heat and injected them into mice. Now, as expected, neither the R nor the S heat-treated forms caused disease. But in the course of these experiments, Griffith made a wholly unexpected discovery that would lead to the identification of DNA as the hereditary material. When Griffith injected mice with both the heat-treated and thus harmless S strain and the alive but nonvirulent R form, the mice developed pneumonia. Even more surprisingly, samples Griffith took from the sick mice showed they harbored living S-form bacteria! To explain this very unexpected result, Griffith postulated that the R strain must have somehow taken up a "transforming principle" from the dead S strain, thus transforming them into the virulent S strain.

In the early decades of the twentieth century, the smart money on the material that encoded Mendel's elementens and caused the transformation of the R strain into the S strain in Griffith's experiments was on the protein component of the bacteria. Other cellular components, including DNA, were considered to be too simple to perform such a complex role. In order to isolate the transforming principle of Griffith, Oswald Avery and his colleagues treated the killed S strains to remove different organic compounds and then tested whether the treated S bacteria could still transform R bacteria into a virulent form. Two experiments were critical. When the S bacteria were treated with enzymes that chew up proteins (proteases), the S bacteria could still transform R bacteria into virulent forms just fine. But when the S bacteria were treated with enzymes that chew up DNA (deoxyribonucleases), the S bacteria could no longer transform. In a 1944 publication, Avery and colleagues concluded that DNA was the transforming principle.[2] This conclusion was definitively confirmed six years later by Alfred Hershey and Martha Chase using a different experimental system focused on viral infection of bacteria. In the two sets of experiments, DNA was confirmed as the organic compound that carries hereditary information.

But how? Unlike the progress from Mendel's work until the experiments of Hershey and Chase, covering nearly an entire century, genetics now moved into a phase of remarkably rapid advance. Only one year after the confirmatory work of Hershey and Chase, the seminal paper of molecular biology was published, suggesting the structure of DNA. This most famous of discoveries was made by two scientists working together in Cambridge, England, beginning in 1951; there an American geneticist, James D. Watson, joined the English physicist and protein structure biologist Francis Crick to focus on the big question of the day—how does DNA encode its information?[3] Before describing their discovery, it is important to emphasize what was already known. Watson and Crick did not discover DNA or even, as we have heard, discover that it is the hereditary material. The molecule that is DNA had been discovered nearly a century earlier by a Swiss chemist, Friedrich Miescher. And Watson and Crick had many of the critical pieces of the puzzle available to them. In addition to X-ray-based images of the structure of DNA, they also had a key insight from the Austrian-American chemist Erwin Chargaff, who was working at the Columbia University Medical School. It had been previously determined that DNA is made up of nucleotides and that each nucleotide includes a sugar component, a phosphate component, and a nitrogenous base component. What Chargaff determined is that the four different bases, A for adenine, T for thymine, C for cytosine, and G for guanine, were not all represented in equal proportions but that the amount of A was always equal to the amount of T and the amount of C was always equal to the amount of G.

Watson and Crick worked aggressively to pull together all the relevant published data and indeed some unpublished data, the use of which led to serious questions about whether the contributions of the only woman directly involved, Rosalind Franklin, were fairly recognized (they certainly seem not to have been).[4] In looking at Dr. Franklin's X-ray images, Watson realized that they suggested a double helix, and Crick realized that the two DNA strands must be running in opposite directions in the complex. They then published their theory in *Nature*, in 1953, with a paper that had undergone no peer review and was only two pages long, including a single hand-drawn figure and one very big idea—that DNA forms a structure of two helical chains coiled around the same axis, with the chain held together by Cs paired to Gs and As to Ts, thus explaining Chargaff's rule.

And in what is perhaps the most audacious understatement in all of biomedical research, the authors noted in the penultimate sentence of their brief paper that "it has not escaped our notice that the specific pairing we have postulated immediately suggests a possible copying mechanism for the genetic material." Watson and Crick realized that the structure itself, with the Cs bonded to the Gs, and the As to the Ts, would permit the information to be copied (for reproduction or to move information within an organism, as we shall see shortly) simply by means of the strands separating and new nucleotides matching with their complementary bases. This stunning understatement, with just one sentence telegraphing the discoveries that would soon be articulated as the Central Dogma of genetics and become the foundation of the golden age of molecular biology, has maintained an iron grip on the imaginations of biologists to this day. When the initial sequencing of the human genome was reported in *Nature* in 2001, the authors explicitly modeled their closing, with perhaps somewhat less justification, on Watson and Crick's famous sentence, noting that "finally, it has not escaped our notice that the more we learn about the human genome, the more there is to explore."

The final piece of our puzzle to set the stage for understanding the chapters to come is to show how the information in DNA is copied between generations (hard enough) and how it is transferred to the machinery that actually builds organisms (even harder!).

Watson and Crick themselves proposed one hypothesis for how it is copied, which was only hinted at in their "understatement of the twentieth century" but developed in a longer follow-up paper only months later. They argued that DNA replication is semiconservative, with the two strands unwinding and being fully copied through complementary base pairing. In this model, the "old" DNA strand would remain intact and would become partnered to a wholly new strand built through base complementarity. Other scientists proposed a conservative hypothesis, in which both strands served as templates for two entirely new strands, and a dispersive hypothesis, which involved breaking up the DNA into small segments and connecting old and new strands. The experiment that resolved the debate was performed by the American geneticists Matt Meselson and Frank Stahl. Meselson had trained with Linus Pauling, the great American chemist who was perhaps Watson and Crick's most formidable competitor in the race to deduce the structure of DNA. And American politics may well have handed the prize to the Cambridge team since the American government prohibited Pauling from traveling to Europe during the critical period leading up to the discovery thanks to his anti-government and anti-war sentiments.

The Meselson-Stahl experiment involved growing the bacterium *E. coli* in a medium that carried a heavier than normal form of nitrogen. As DNA replicates, it takes up the heavier nitrogen. The result is that different weights for DNA are expected based on which model of DNA replication is the correct one. These experiments, sometimes called the most beautiful in biology, definitively ruled out the conservative and the dispersive hypotheses and confirmed the Watson and Crick hypothesis of semiconservative replication. An

—

enzyme related to one that actually did the copying (one form of DNA polymerase) was discovered by Arthur Kornberg, for which he almost immediately won the Nobel Prize in 1959.

This leads us finally to how the genetic code is read. Proteins were what scientists had bet on as the hereditary material earlier, as proteins are the complex organic compounds found everywhere in organisms and responsible for most of what organisms do and how they look. With DNA confirmed as the hereditary material, there had to be a way for the information in DNA to be transferred to proteins in order to build organisms. It remained clear that proteins were all-important to cells and organisms, but they must be following the blueprint specified in the DNA. But how is this information transferred from DNA into proteins?

The key discoveries required for the eventual resolution of the genetic code occurred in a complex fashion in the decades surrounding the discovery of the structure of DNA in a sequence not easily discerned from the published record. One key element, however, was provided by work beginning in the 1940s that showed that a DNA-like substance, ribonucleic acid (RNA), was abundant in the cytoplasm of cells. One other key element was the simple quantitative consideration of how a sequence of DNA bases might code for amino acids. It was known that proteins are made up of twenty amino acids (the exact set of twenty was still being debated). This obviously rules out a one-to-one mapping between DNA bases and amino acids, since four DNA bases can encode only four amino acids in a mapping of one base to one amino acid. So the "words" encoded in DNA spelling specific amino acids (which we now call codons) evidently involve more than one DNA base, but how many more? Perhaps the first to specify this question in clear mathematical form was the cosmologist George Gamow. In a handwritten letter to Watson and Crick in 1953, he articulated the challenge of divining a genetic code to translate a sequence of DNA bases into a sequence of amino acids in proteins. Noting that if the DNA words involved only two bases, there would be 4 × 4 possible words, or sixteen different amino acids (AA, AT, AC, AG; CA, CT, CC, CG; TA, TT, TC, TG; GA, GT, GC, GG), and not enough words. Gamow reasoned that the DNA word therefore involves three DNA bases, which would lead to sixty-four different three-letter words (4 × 4 × 4), comfortably allowing representation of all amino acids with some potential for redundancy.

As is often the case in biology, key experiments are built around novel conceptual frameworks. The more accurate the conceptual framework, the more informative the experiments. A key contribution to the conceptual framework came, fittingly enough, from Crick himself. Noting that DNA is concentrated in the nucleus of a cell, and both RNA and proteins are abundant in the cytoplasm, Crick needed a way to get information from the DNA in the nucleus into the cytoplasm to serve as a template for linking amino acids into proteins. From this was born what is sometimes called the Central Dogma lecture. Crick argued that there was a specialized RNA molecule that would serve as a messenger moving the information in DNA into the cytoplasm, where this messenger RNA would interact with other specialized RNAs whose job it was to deliver the correct amino acids. This process would constitute a one-directional flow of information from the DNA in the nucleus into the sequence of amino acids in proteins, but never back again. While informed by some knowledge of potential players in the flow of information, Crick's Central Dogma was above all a statement about the flow of information from genes into protein sequences. With this framework in place, the race was then on to confirm the length of DNA

—

2nd letter

		U		C		A		G	
1st letter	**U**	UUU UUC	Phe	UCU UCC	Ser	UAU UAC	Tyr	UGU UGC	Cys
		UUA UUG	Leu	UCA UCG		UAA UAG	Stop	UGA UGG	Stop Trp
	C	CUU CUC	Leu	CCU CCC	Pro	CAU CAC	His	CGU CGC	Arg
		CUA CUG		CCA CCG		CAA CAG	Gln	CGA CGG	
	A	AUU AUC	Ile	ACU ACC	Thr	AAU AAC	Asn	AGU AGC	Ser
		AUA		ACA		AAA AAG	Lys	AGA AGG	Arg
		AUG	Met	ACG					
	G	GUU GUC	Val	GCU GCC	Ala	GAU GAC	Asp	GGU GGC	Gly
		GUA GUG		GCA GCG		GAA GAG	Glu	GGA GGG	

Figure 2.3. A representation of the genetic code showing the amino acids encoded by each of the possibly sixty-four nucleotide codons. (Drawn by Ryan Dhindsa, Goldstein Lab)

words, to characterize these elusive RNA species, and, above all, to learn to read the code of life.

Crick and a number of colleagues performed the first experiments to confirm Gamow's conjecture that the DNA words were indeed made up of three bases. They did this by introducing what we call frameshift mutations, deletions or additions of DNA that shift the reading frame. They found that a single base pair deletion would disrupt the protein that was synthesized but that three single base deletions would restore the reading frame. Hence, the words were triplets. From here, scientists constructed artificial RNA sequences and laboriously determined the resulting proteins that were made to build up the complete genetic code. This work can be seen

—

as providing the precise dictionary permitting the flow of information specified by Crick's Central Dogma. Specifically, we can view the genetic code as providing an information map that translates a combination of four different elements (U, C, A, and G, with U replacing T in RNA) into one of the amino acids. With the word size determined as three, the full mapping can be determined and presented in various formats (such as that shown in figure 2.3). The way to read the map is simply to identify the three-letter word and determine which amino acid it is linked to. Thus, in the first half of the first box, UUU and UUC code for phenylalanine, for example. At long last, scientists had cracked the code of life. And in fact, just by studying a figure such as this, you too can read the book of DNA and see how any gene in the human genome encodes for specific amino acids and the proteins they create.

Chapter 3

THE NATURE OF HUMAN GENETIC VARIATION

Advances in our understanding of genetics and evolution over the past approximately 150 years rival any insights about the natural world that human imagination and careful investigations have ever conjured. As we have seen, following Mendel's remarkable early experiments with garden peas and Darwin's explanation of life on earth as evolving through the natural selection of individuals carrying inherited differences, we now understand the underlying core principles of both genetics and evolution. We now know that there are four thousand genes that cause serious human diseases, following exactly the rules of inheritance that Mendel painstakingly described, and that collectively afflict hundreds of millions of humans with often devastating and incurable diseases.[1] And no serious practitioner in any branch of the life sciences today can reasonably doubt that the adaptations we see throughout the living world today arose through natural selection.

As we also learned earlier, today we not only know that DNA is the material substance of heredity but also know exactly how DNA encodes its information and how it is copied, almost, but not quite perfectly, from generation to generation, with imperfections introduced and carried forward by each and every one of us in the form of new mutations. These mutations are exactly the heritable differences that Darwin postulated that natural selection acts on. These are uncontested triumphs of reductionist modern science accurately explaining core principles of both genetics and evolution applicable to all living things and rightfully earning a number of grand labels. These include the Modern Synthesis, which unites Mendelian inheritance and Darwin's theory of evolution by natural selection, and the Central Dogma, which describes the flow of information from DNA through RNA to the proteins responsible for the diversity of organisms on earth, using the genetic code that we, and now you, have learned to read. Coupling all of this with the triumphs of genetics written about on a daily basis in the world's press, from mapping the human genome to identifying the single chemical unit among three billion responsible for a child's devastating disease, you could be forgiven for thinking that we now understand the really important questions about evolution and genetics. How did cheetahs come to run so fast and blue whales come to traverse the oceans and fill them with their songs, and how did humans develop the intelligence necessary to ask and seek the answers to such questions? Unfortunately, nothing could be further from the truth. Although the trees in the genetic forest have been described, the forest remains almost entirely inscrutable to us.

This claim of near complete ignorance concerning answers to the big questions about genetics and evolution will not only be surprising to those outside the field but no doubt irritating to some of the more bullish professional genomicists. But it is in fact quite easy to see how little we really know by looking at the history of evolutionary genetics and the big questions the field wrestled with over the past hundred years. Revisiting this history is especially critical now, sitting, as we do, on the precipice of sculpting the genomes of future generations. As we embark on this process of selecting the genomic characteristics of our children, we would do well to fully understand how much we still do not know. Unfortunately, we find ourselves in a situation in which our technical skill dramatically outpaces our understanding of the consequences of what we will be able to do all too soon now.

To try to tell the story of how little we really know we turn to James F. Crow, one of the giants of twentieth-century genetics, who over a career of seventy years was directly involved in nearly all of the major questions and debates concerning population and human genetics. Beyond his scientific contributions, Professor Crow also differentiated himself as perhaps the kindest and most gracious of all the major figures in genetics. To those fortunate enough to know him, he seemed transported into genetics from an entirely different era or world. Despite a nearly encyclopedic knowledge of experimental genetics and a comfortable command of its mathematical descriptions, Crow never had an unkind word to say about his colleagues, was prepared to change his mind about his own theories when new experiments challenged those theories, and dedicated inordinate time and energy to helping students and colleagues better understand their own work. I had the benefit of experiencing this firsthand when Crow unexpectedly reviewed my first sole-authored paper, helping to simplify what had previously been an overly complicated interpretation of a relatively simple mathematical model.[2] I later had the opportunity to meet Professor Crow in a meeting he organized that focused on a topic related to that first paper. In a field often dominated by personal rivalries and self-promotion, Crow embodied a way to practice science competently, but graciously, remaining interested above all in what is true.

Writing near the end of his life, in 2008, Crow reviewed the controversies that had consumed him and evolutionary geneticists throughout the twentieth century. He described geneticists as seeking general principles embodied in four key, and partially overlapping, controversies: (1) the shifting balance theory of Sewall Wright; (2) the causes of heterosis, in which combinations of different strains of organisms are observed to be more fit than their parental strains; (3) the classical versus balanced school of genetic variation; and (4) the neutral theory of molecular evolution.

These debates need to be revisited and understood not only for their historical importance but to make clear that these key questions about the nature of genetic variation, human genetic variation in particular, have yet to be fully answered. Instead, the field simply moved on to things it could answer and do, as Crow himself noted was already happening in 2008. Unfortunately, this resulting ignorance means that as we begin manipulating the genomes of future generations, we do so in a state of much greater ignorance than is generally acknowledged by professionals or appreciated by the general public. To fully appreciate this state of ignorance, it is important to understand the controversies that Crow highlighted and what we do and, more important, what we still do not know, about the resolution of those controversies. We will now review each of these controversies highlighted by Crow and then consider what their resolution, or lack thereof, tells us about how people will, or should, use the newly available technologies in human reproductive genomics. The astute reader will recognize that the controversies are in fact related to one another, which means that answers to one of them

—

would likely inform the others, but that these underlying and uni-fying answers remain elusive. Although some of the material I am going to cover will be a little detailed for a non-specialist reader, after reviewing the controversies we will return to the implications for reproductive genomics, so please trust me that it is worth the effort to follow the historical arguments.

Controversy 1: The Shifting Balance Theory

The quantitative description of how Mendelian inheritance inter-acts with both natural selection on allelic forms and random fluc-tuations due to smaller than infinite population sizes is a branch of biology that came to be referred to as population genetics. Before the Human Genome Project and related advances drew population geneticists into medical genetics and other very applied efforts, pop-ulation genetics was quite a rarefied discipline, with very few prac-titioners. And certainly no money was dedicated to the field, unlike today, for better and for worse. Indeed, the origins of the mathe-matical basis of classical population genetics can be traced almost entirely to only three scientists — Ronald Fisher, Sewall Wright, and J.B.S. Haldane.[3]

Conceptually, Fisher considered the problem a straightforward process in which mutation introduces variants that are, on average, less fit than the average of the population (usually) and occasionally variants that are more fit, and these then increase in frequency, leading to an increase in overall fitness of the population and thus, eventually, striking new adaptations like wings for flight. Certainly the mathematics works if one simply assigns levels of fitness to alleles and allows the recursion equations to carry those new alleles on to fixation in the population. The adaptations we see all around us are then the results of a great many such fixation processes of new advantageous alleles across many genes, while selection ruthlessly eliminates the nasty alleles that appear spontaneously in each generation.

Using the same kinds of considerations, Haldane raised a concern about how much adaptation this process would in reality be capable of generating. His concern emphasized that positive selection favoring a good allele comes at the expense of selective mortality working against the unfavored allele. Thus, for any of the sweeps that Fisher imagined, you would need to have selective death (or failure to reproduce) for some proportion of the population to drive the increase of the more favored allelic form. Haldane referred to this as the cost of selection. His mathematical characterization led him to postulate that the rate of evolution would therefore be constrained by the cost associated with natural selection. Subsequently, other mathematical treatments have shown that the constraints imposed by the cost of selection are highly dependent on how one assumes that selection acts. In particular, the possibility that selection could eliminate sets of less fit alleles together would allow evolution to move much faster at a given cost of selection. That is, in terms of the number of positive alleles that sweep through the population, you could end up with more positive alleles selected with the same amount of selective death if selection acts against groups of alleles together. Under this model, the cost of selection does not appear as a barrier to rapid evolutionary change.

Sewall Wright, however, believed that there was a more fundamental problem to address. His own work in experimental genetics, in particular with coat color in guinea pigs, convinced him that evolutionary geneticists had a paradox to explain. He argued, quite reasonably, that complex adaptations must depend on having the right alleles at multiple different genes. Cheetahs do not run seventy-five miles an hour because a new allele has appeared in the "run fast" gene, and camels cannot cover a hundred miles of parched desert with no water because of a new allele in the "hump fat storage" gene. Although we do not know all the genetic changes responsible for these adaptations, we do know that they depend on changes throughout the genomes of cheetahs and camels that, in turn, change their physiologies in ways that allow brief bursts of exceptional speed and remarkable endurance, respectively. This, we know, is true of most complex adaptations. With some rare exceptions, changes to a single gene are capable of entirely breaking complex adaptations but are not capable of building them in the first place. A clear illustration of this phenomenon is human intelligence. We know for a number of reasons that no single gene is responsible for the cognitive superiority of humans over other primates. But we also know that there are up to about a thousand genes that, when mutated, dramatically reduce the intellectual capabilities of people, including, for example, making them incapable of speech.

Considering the need for genetic changes to work together to create complex adaptations, Sewall Wright felt that evolutionists had a major problem on their hands. His life work would probably be fairly described as beginning, like Albert Einstein's, with a *Gedankenexperiment,* or thought experiment. That is, instead of performing an experiment, imagine an experiment conceptually to see what might happen. Wright imagined a population that currently has a nice set of adaptations and asked, How does that population evolve an entirely new adaptation? The problem is that, for the set of current adaptations, alleles have all evolved to work together across multiple genes. If a new mutation appears that would be beneficial in a new adaptation, it would be advantageous alongside a bunch of alleles supporting that new adaptation, whereas it would be useless or, more likely, harmful in the context of the current adaptations that a population is enjoying. Put in Wright's language, there is a co-adapted gene complex predominant in the population permitting the current adaptation, and an allele that might be part of a different co-adapted gene complex would not be at a selective advantage with the current genetic background.

So how in the world does a population move from one co-adapted gene complex to another or from one suite of adaptations to another? To make this concrete, in caricature form, consider the cetaceans, the mammalian order that includes baleen, toothed whales, and dolphins. The analysis of genetic, morphological, and paleontological evidence makes clear that cetaceans evolved from land-based animals related to modern-day even-toed ungulates (deer, hippos, sheep, camels, and the like), with the most recent ancestor of both cetaceans and land-based even-toed ungulates being a hippopotamus-like creature that lived some fifty million years ago on the Asian subcontinent. How did genetic variants appear that turned a hippopotamus into a blue whale? The vast majority of genetic changes that distinguish a blue whale from a hippo would have been rather unfortunate for the individual hippo still trying to make her living on land.

To get around this fundamental problem, Sewall Wright dreamed up the shifting balance theory. He imagined that populations of species are partially isolated from one another. And within different parts of the larger population, there are relatively few individuals allowing random changes in their genetic composition due to genetic drift. Therefore, the large population has many subpopulations, only partially connected by limited genetic exchange, that are free to randomly explore new genetic combinations. Occasionally, some part of the larger population hits on a winning new suite of alleles across multiple genes. And once this occurs, the carriers of this new, fortuitous combination across multiple genes are at an advantage, and the new co-adapted complex of genetic changes then sweeps through the broader population. Hence the name "shifting balance." At a certain point in time, one part of the broader population has the best combination, and then at some other point in time, another part does. Over a sixty-year period, Wright published papers outlining mathematical details of his theory, and in so doing he developed many of the fundamental concepts of theoretical population genetics. But the theory, although it was appealing to a few highly influential figures throughout the twentieth century, most notably the paleontologist George Gaylord Simpson and the evolutionary geneticist Theodosius Dobzhansky, is all but ignored now by working geneticists. Indeed, personal experience confirms that today a professor asking graduate students in genetics their views about the shifting balance theory will, more often than not, elicit a slightly nervous blank stare.

In fact, there are sound mathematical reasons to be skeptical about the shifting balance theory, above all related to the central question of the exact nature of the population structure that would permit it to operate. In order for subpopulations of a broader population to be sufficiently isolated to allow independent random "exploration" of beneficial combinations of alleles across genes, mixing throughout the population has to be sufficiently limited. On the other hand, in order for successful new co-adapted gene complexes to take over the broader population, genetic exchange throughout the population would need to be at a sufficiently high level to allow beneficial alleles to get from one of the subpopulations to the rest of the population. Careful mathematical modeling suggests that while there is indeed a sweet spot of local mating and limited gene flow across groups that allows both processes to operate, the parameters permitting this represent very specialized conditions. Generally speaking, scientists are always uncomfortable when the parameters need to be just right for a theory to work. The fear, of course, is that those just-right conditions will rarely obtain in the real world.

So where does this leave us? My own view is that there is little if any experimental evidence that supports Wright's life work. And yet it probably remains the most serious effort to date to solve this vexing puzzle of how evolution can transform a population with one set of adaptations into a population with an entirely new complex of adaptations. Despite the clear limitations of the Wright theory (sensitive dependence on precise parameters of genetic mixing throughout the population and the absence of any direct experimental evaluation, let alone validation), it remains the strongest candidate to provide an explanation conceptually. Wright's theory is particularly compelling if we assume that adaptations do indeed depend on networks of genes that have alleles that must work together, as seems certain. And if we return to our central question about engineering human genomes, we must face the reality that we will be doing so in almost complete ignorance of how selection shaped our genomes to create the human attributes we most care about. The puzzle that Sewall Wright spent his life addressing does not seem likely to be solved anytime soon — almost certainly not before we have the technical ability to choose the C, T, A, and Gs we would like our children to carry.

—

Controversy 2: Heterosis, Dominance Versus Overdominance

"Heterosis" refers to superior characteristics of organisms that result from the combination of parents that are derived from different strains, or "lines," of the species. The reality of heterosis in the natural world is uncontested, and evidence of it can be found in what most of us eat almost every day. The most dramatic illustration is hybrid corn. Experiments informed by the developing science of genetics in the early twentieth century showed that producing corn seeds from different inbred corn strains led to hybrid plants with much greater yields than those of the parents. In the decades following this discovery, hybrid corn became the dominant form and led to a dramatic increase in productivity, arguably providing, before the modern genomics era, the most commercially important application of the science of genetics. But why were the hybrid corn plants much more productive than the parents?

The science of genetics offers two distinct explanations. The first is dominance. As you will recall, in diploid organisms there are two alleles present for each gene, one inherited from the male and one from the female parent. Under the dominance explanation, the hybrid does better because single copies of harmful alleles (inherited from only one parent) are silenced by the dominant "normal" alleles from the other parent. Under this hypothesis, each parental strain has some number of harmful alleles present, and the effects of these harmful alleles are masked by the normal or wild type allele of the other strains. Since the hybrid has most of these harmful alleles masked, it does better.

The alternative explanation is what geneticists call overdominance. Overdominance occurs when individuals carrying two different forms of an allele do better than individuals that have two copies of the same allele (homozygotes). For example, if we denoted the alleles at gene A as A1 and A2, individuals that are A1A2 heterozygotes are more fit (with either a survival or a reproductive advantage) than individuals that are either A1A1 or A2A2 homozygotes. Although this is a simple idea, there are strikingly few documented examples of heterozygote advantage. The most famous one in humans relates to sickle cell anemia. Sickle cell disease results from mutations in the HBB gene, which encodes part of the hemoglobin protein, responsible for transporting oxygen from the lungs (or the gills if you are a fish) to the rest of your body. Some people carry a mutation that leads to an abnormal form of hemoglobin called hemoglobin S (there are also other mutant forms we are not considering here). When a person carries only one copy of hemoglobin S, he is protected against the *Plasmodium* parasite that causes malaria, while people that carry two copies of hemoglobin S suffer from sickle cell disease. Although the symptoms of sickle cell disease are variable, patients have characteristically sickle-shaped red blood cells, which leads to the cells' premature degradation, resulting in anemia and other effects. In environments with high levels of exposure to the *Plasmodium* parasite (carried by *Anopheles* mosquitos), the heterozygotes (with one hemoglobin S allele and one normal allele) do better than either those with two normal alleles or those with two hemoglobin S alleles, providing a clear example of heterozygote advantage. In the absence of malaria, however, the best genotype to have is clearly two normal copies, showing the relevance of the environment to genetic fitness. The sickle cell example and a very few others like it show that heterozygote advantage can occur, but the paucity of these examples makes it very unclear how common it is, and also makes it unclear whether the superiority of hybrid corn and related examples is due to either dominance or overdominance.

—

69

During the middle of the twentieth century numerous experiments were conducted in an attempt to resolve the issue, with James Crow himself first arguing that population genetic estimates of the genetic load of harmful mutations suggested that silencing those mutations would be insufficient to explain the quantitative benefits of heterosis, thus supporting the overdominance hypothesis. Crow, however, later changed his mind about his own interpretation in the face of evidence that recessive mutations are rarely entirely recessive, meaning that they have a lower prevalence in the population because selection can act against them even in heterozygote forms. This means that there is a greater genetic load than Crow had assumed, since selection was acting against mutations even in heterozygote form. New estimates of the mutation rate further weakened the argument, leading Crow and others to believe that dominance could not, in fact, be ruled out as an explanation and that heterozygote advantage may not in fact be needed as an explanation for hybrid vigor. And where is this debate left today? More or less where Crow left it. We do not have solid evidence telling us which hypothesis is right, and it is not even a topic that many geneticists work on now.

So how does this affect our thinking about reproductive genomic engineering? Unfortunately, it makes it overwhelmingly clear that we, in fact, do not know what we are doing. To make this point clear, let us return to the thought experiment we introduced earlier describing possible reproductive genome engineering strategies and their potential consequences. The introduction raised the concept of the "common-variant human." In this thought experiment, we imagined that upon sequencing of parent genomes all rare gene variants are identified and the embryos designed for transfer have each of these rare variants replaced with what is the most common variant in the population. This will, of course, lead to a dramatic reduction of heterozygosity or, depending on exactly how the experiment is implemented, the elimination of heterozygosity. Let us now imagine that the overdominance hypothesis is right. This would mean that having only one form present at many genes in the human genome would result in a person with a much lower than average level of fitness. In this case, creating human embryos with only common alleles would result in individuals that are much less fit than average in a broad range of currently unpredictable ways. On the other hand, if the dominance theory is right, for most genes the common form will be the normal "dominant" form and the creation of embryos with only the common allelic forms will very likely result in humans that are more fit than average. There are, of course, nuances to this prediction, which will be best considered after we have reviewed the classical and balance schools, considered next.

Controversy 3: The Classical Versus the Balance Hypotheses of Genetic Variation

As just outlined, the heterosis debate related to the causes of the improved fitness of hybrids. The question was not whether most genes showed heterozygote advantage but rather whether heterozygote advantage (that is, the presence of both allelic forms at a gene is advantageous) is the primary driver of hybrid vigor, as opposed to the silencing of disadvantageous alleles by the dominant form. Population genetics models show that heterozygote advantage at a rather small proportion of genes could still be primarily responsible for hybrid vigor. The classical and balanced schools of thought can be viewed as the generalized form of this same debate: What happens at most genes in most populations? Are heterozy-

—

71

gotes generally more fit than homozygotes? The originator and principal advocate of the view that heterozygotes are often at an advantage was Theodosius Dobzhansky, a Ukrainian-born geneticist who spent his career in the United States. More than any single figure, Dobzhansky was responsible for developing a science of ecological genetics, studying how and why genetic variations in the fruit fly changed in frequency in natural populations. Dobzhansky's work convinced him that selection favored the presence of two different allelic forms in many or even most genes, a form of selection referred to as balanced.

There could be many different reasons that having two forms of a gene might be beneficial to an individual, but to see how it could work, consider two simple examples. If a gene encodes an enzyme that works only in a finite temperature range and if an organism experiences a broader range of temperatures, it is easy to imagine that having one form of the enzyme optimal at one temperature and another form optimal at a different temperature might be preferred by selection, compared with having only an enzymatic form that is optimal at a single temperature. Or, if a gene encodes a protein that presents specific foreign proteins to train immune cells to attack cells that are infected with viruses (thus full of foreign protein fragments), it is not hard to imagine that having two forms able to present a broader range of foreign protein fragments might be better than only one. Dobzhansky argued that this kind of selection is rampant and, as a consequence, the level of fitness is higher for most genes when there are two allelic forms present instead of one.

Dobzhansky's views were most keenly contested by a contemporary geneticist also working with fruit flies, Hermann Muller. Muller's early work focused on the study of lethal mutations in flies, and in the process of working to introduce such mutations, Muller characterized the mutagenic effects of radiation, earning him the Nobel Prize in Physiology or Medicine in 1946. Likely due to his study of laboratory-induced mutations, Muller held the view that most genes had a "normal" or wild type allele and that mutation introduces harmful variants into the population at a steady rate. This led him to conclude that most genes that show variation carry a good and a bad allele. This view, sometimes called the classical view or school, implicitly posits that when positive selection acts, the superior allele introduced by mutation relatively quickly overtakes the population, so much of the variation that you observe at any given time is due to the difference between the good wild-type allele and low-frequency harmful alleles constantly introduced by mutation, since beneficial mutations are far less common than harmful ones. When you look at differences among people, the balanced view would be that different people have different combinations of beneficial alleles, while the classical view would be that different people are carrying different, and overall different loads, of harmful alleles.

We know today that both kinds of variation exist. In some settings, having two alleles is beneficial — for example, as we discussed, in the hemoglobin HBB gene in the presence of malaria — and we know that there are certain immune-related genes that show similar effects. We also know that new mutations, in every generation, introduce clearly deleterious effects that conform neatly to the Mullerian view. These are indeed precisely the kinds of mutations we most often identify as responsible for Mendelian diseases, as will be seen in the next chapter. What we still do not know is which view is more correct when we compare people who do not have devastating diseases. Even without a genetic disease, we all still differ from one another in our height, weight, longevity, types of intelligence, other

—

behaviors, and in the particular complex disease that most of us will eventually manifest. Although we can say that many of the genetically simpler diseases conform to a Mullerian view, we cannot say which school has the best claim to the causes of non-disease-related differences among people. Nor can we say which school has the best claim to explaining the genetic causation of more complex common diseases.

Controversy 4: The Neutral Theory of Evolution

The neutral theory of evolution was developed to explain an apparent paradox. The comparisons of different species suggested a rate of molecular evolution that seemed, to some, to be too high to be consistent with Haldane's cost of selection, as we saw previously. To overcome this paradox, Motoo Kimura, a Japanese geneticist who worked with James Crow through much of his career, introduced the idea that most genetic changes are selectively neutral, meaning that natural selection cannot act on them. Therefore, the rate of molecular evolution depends on the mutation rate, not the nature of selection. And Kimura argued that the rough constancy of molecular rates of change across species and across evolutionary time (the so-called molecular clock) could be explained only by neutral molecular evolution, as selection could not reasonably be expected to produce roughly constant rates of change. This claim was in fact challenged by various geneticists who showed that changing selection pressures could yield roughly constant rates of change, while others questioned how constant the molecular clock really is. Like the previous controversy, this debate has largely subsided without a clear resolution, but most biologists today probably sub-

—

scribe to the idea that large parts of the genomes of most species evolve nearly neutrally, with small proportions of genomes, including much of the portions of genomes that encode protein, under strong selection.

Whatever the reason, it is clear that most parts of genomes do follow a rough molecular clock. That is to say, for most comparisons among species, the amount of divergence in the sequences of their genomes correlates with the evolutionary distance between them. Here evolutionary distance means the amount of time at which some common ancestral form separated into two distinct groups that became reproductively isolated from each other and eventually became separate species that are unable to reproduce with one another. For example, the closest living relative of humans is the chimpanzee, and humans and chimps are thought to have separated from each other at least five million years ago, and perhaps as much as ten or so million years ago. In fact, many such speciation events do not happen cleanly, and there is often a long period of interbreeding following partial separation that complicates the estimation of separation times. Indeed, this has been postulated for the human-chimp split itself and has been further postulated for many of the further partial separations among lineages leading to modern humans and other lineages. Perhaps the most famous example of such interbreeding following partial separation among species is due to the work of Svante Pääbo, David Reich, and colleagues, who used sequence data from Neanderthal specimens to argue that there was a period of interbreeding following the separation of modern humans and Neanderthals that has left most humans today with a smattering of Neanderthal genes. Absent such interspecific amorous complications, however, the expectation would be that the diver-

—

75

gence of the genomes of, for example, chimps and humans would reflect the time since the populations leading to each modern form separated from one another.

Regardless of how quantitatively accurate a molecular clock is, we do know that when the genomes of different species are compared, many of those genomes show a quantitative degree of change that correlates fairly well with the time that fossil evidence would have suggested that they separated from one another. Beyond questions about post-separation hybridization, many other questions about the molecular clock have emerged, including one about whether generation lengths influence the clock beyond absolute separation times. Overall, while the clock has been referred to as a sloppy clock, there is no doubt that it ticks, and no doubt that it ticks at different rates for different parts of the genome. Parts of the genome that are critical to survival clearly tick at a much slower rate, on average, than parts of the genome that are less critical to survival. Indeed, as we see more fully in chapter 4, the degree of conservation among species is now routinely used, among other metrics, to prioritize the human genome in terms of regions most likely to cause disease. Regions that are relatively resistant to evolutionary change across disparate species are likely to be essential to function. Like the other controversies discussed, however, the reason for a rough constancy in the molecular clock remains unknown. If the majority of the human genome really is selectively neutral, it does not matter at all what particular bases you have at most of the three billion sites in your genome. This remains entirely unclear, however. We know from experiments in model organisms and from human disease genetics that some parts of the genome are much more important than others. For example, if someone told me I was obligated to give up a few base pairs out of my genome, I would most certainly

—

choose a region that does not make protein over a region that does. We know that most changes outside of protein-coding regions are not incompatible with healthy life, while many changes within genes are incompatible with healthy life. But this does not confirm that most of the genome is under no selection pressure. The proportion of the genome that truly follows neutral evolution remains, much like the other controversies, simply unknown.

In keeping with my focus specifically on the question of human reproductive design, I have presented what we know about human genetic variation without much consideration of the broader evolutionary context. Similarly, in the previous chapter we discussed Mendel in detail, along with the discovery of the material basis of heredity, but nothing about Darwin's theories. I should, however, emphasize that while I have not presented the conceptual development of evolutionary theory explicitly, my presentation of genetics is predicated on our current understanding of evolution and specifically how genetics and evolution are united in what has been called the "Modern Synthesis." Defining the Modern Synthesis is a challenge even at length, let alone briefly. This is probably due to the fact that it is more of an ongoing synthesis of how disparate branches of science inform our understanding of the evolution of life on earth. Especially given the current lamentable challenges to well-established scientific principles, it is important to provide at least a superficial rendering of the Modern Synthesis.

When Mendel's rules were rediscovered at the beginning of the twentieth century, evolution and genetics were largely separate disciplines. Evolution itself was increasingly accepted by scientists, but how it operated remained mysterious, and ideas surrounding evolution remained largely independent of the burgeoning science of genetics. This changed with work beginning in the 1920s that de-

—

veloped a conceptual framework for how to unite genetics and evolution by natural selection, in what was coined as the Modern Synthesis in 1942 by the evolutionist Julian Huxley. The family name is likely familiar because of multiple prominent members, including Julian's grandfather, Thomas Henry Huxley, known as Darwin's bulldog; his brother Aldous Huxley, the author of *Brave New World;* and his half-brother Andrew Huxley, who won the Nobel Prize in Physiology or Medicine for his work on action potentials of nerves.

From my perspective, the Modern Synthesis can be seen most simply as a way to unite the Mendelian view of genetics with observations of both natural historians and paleontologists into a coherent theory of evolutionary change over time. From this perspective, we can encapsulate the Modern Synthesis as viewing evolutionary change in species over time as the changes in the frequencies of genetic variants over time in populations, with consequent morphological consequences in populations that can be discerned in the fossil record. Although there have been many reformulations of this synthesis over time, working geneticists and evolutionary biologists do not question its core principles. This is perhaps particularly important to emphasize today with modern attempts to again separate evolution and genetics, as, for example, in some of the beliefs of Creation Science. Working evolutionary biologists and geneticists may have plenty to argue about, such as whether individual genes are really the targets of natural selection and how important cultural evolution is relative to genetic evolution, but these arguments are really all about nuance. I have never met a contributing geneticist or an evolutionary biologist who in any way doubts the core tenets of the Modern Synthesis and its description of the evolution of life on earth.

—

What the Unresolved Controversies
Tell Us About Genome Engineering

The main point of reviewing these signature and almost entirely unresolved controversies is to emphasize how little we really know as we enter a new age that will inevitably involve large numbers of individuals choosing the genomic characteristics of their children. Since these controversies show no real signs of being resolved imminently, it seems all too certain, for better and likely more often for worse, that they will eventually be resolved by the relevant experiments being performed *in people.* I am in no way arguing that these experiments should be done, but rather simply noting that it is likely they will be done, and that they will likely be done from a position of substantial ignorance, as I hope this chapter makes clear.

To illustrate this point as starkly as possible, let us return to our own thought experiment of the common-variant human, now considered in the context of controversy 3. Let us assume that Muller is essentially right and that most differences among people are due to their carrying variable burdens of rare harmful mutations. A couple could use technologies that will most certainly be deployable in the decades to come, as outlined in chapter 5, to eliminate every rare variant in the genomes of their children. Under the Mullerian view of the genetic world, they would find themselves with children who were above the average human condition in virtually all characteristics. How far above the average is a question we return to, but do not answer, in chapter 5. But under the Mullerian view, they would be well above average. Under the balanced view, however, they would be below average in virtually all characteristics. A question that societies may need to confront before long is whether we really want to resolve this debate by generating engineered human beings to find out what they are like.

—

79

Chapter 4

DNA AND HUMAN DISEASE

In order to know exactly what parents might ultimately choose to adjust in the genomes of their children, the starting point is an understanding of what we know about how genetic variation influences human health. But beyond the connections between genetics and health, it is important to also understand how those connections were derived. Only in this way can parents judge for themselves the expected consequences of potential genomic alterations in the genomes of their children. And, critically, only in this way can parents know the extent of uncertainty in some of those expected consequences.

I start with the first important consequences of the Human Genome Project, which resulted in a greatly increased ability to track down the causes of Mendelian diseases simply by looking at the pattern of inheritance of disease in families and testing landmark DNA marker sites throughout the genome. The co-inheritance of these marker sites and diseases then pinpointed the region of the genome carrying the responsible mutation. I should note here that I will use terminology a little different from what is sometimes seen

—

in popular accounts of human genetics. In nontechnical writing, it is not uncommon to find references to the "cystic fibrosis gene" or the "sickle cell anemia gene." And the convenience of such wording even slips into technical writing. Where grammatically workable, I have avoided this shorthand. This is one of the shortcuts that leads to a misrepresentation of genetics. Many and perhaps most genes in the human genome do a great many things. A gene might carry mutations that cause epilepsy, but it might also carry mutations that cause autism, schizophrenia, or kidney disease. And many of the genes that we know carry mutations that cause specific diseases are likely to be involved in a great many functions that we know nothing about and are unrelated to the disease. In this context, it is much more accurate to refer to the mutations causing a specific disease, as opposed to referring to the gene for that disease. Following discussion of efforts to track down disease-causing mutations through co-inheritance of diseases and markers in families, I describe the first systematic efforts to relate genetic variation throughout the genome to the risk of common diseases through genome-wide association studies. Finally I turn to the very current era of sequencing entire genomes to find the causes of disease and ultimately to find genetic contributions to how we look, act, and feel.

One in Three Billion: Finding the Causes of Mendelian Diseases

The first critical contribution of the Human Genome Project was in developing detailed physical and genetic maps of the human genome. These efforts amounted to identifying key landmark sequences that are distributed throughout the genome and that mark off small regions spaced at regular intervals based on either the simple base

—

pairs separating those landmarks (for a physical map) or on the recombination rate separating the markers (for a genetic map). The very first genetic maps were defined based on the observation of mutations in the fruit fly, as described in chapter 2. More than a half-century later, these same kinds of maps proved critical to the remarkable phase of human disease-gene discovery that began in the late 1980s and continued for about twenty years. There are thousands of genetic diseases that are known to run in families in a pattern of inheritance that follows Mendel's rules. Despite having long understood that the causes of these diseases follow the same rules of inheritance that had been described in peas a century ago, until the late 1980s we had no way to systematically track down the precise genetic causes of these diseases. In 1980, however, the geneticist David Botstein and his colleagues articulated a vision of how the responsible mutations and genes could be identified.[1] To appreciate the importance of Botstein's work, we need to go back a further eighty years, to the English physician Archibald Garrod, who is generally considered the founder of medical genetics.

At the end of the nineteenth century, Dr. Garrod was working at the Great Ormond Street Hospital for Children, which is still one of the world's leading children's hospitals for both care and research. While working there, Dr. Garrod developed a particular interest in urine, both normal and abnormal, rightly viewing it as a non-invasive guide to chemical reactions at work in the body. In 1897, a mother came to the hospital to report on the blackened diapers of her child. Dr. Garrod subsequently identified a number of similarly affected families and discussed his results with William Bateson, the first zoologist to champion Mendelian inheritance, having closely followed the work in plants that had caused Mendel's rules to be rediscovered. Together, Bateson and Garrod recognized that the con-

—

dition Garrod had identified, alkaptonuria, followed a recessive pattern of inheritance. This and a few other examples led Dr. Garrod to develop the concept of human chemical individuality and to the systematic study of inborn errors of metabolism.

The terminology of inborn errors of metabolism is still used today to describe a broad class of Mendelian diseases. Beyond recognizing that genetic defects can create chemical individuality among humans, the team of Garrod and Bateson also developed a philosophy that has been at the heart of medical genetics ever since. Writing in 1908, Dr. Bateson advised: "If I may throw out a word of counsel to beginners, it is this: Treasure your exceptions!" And this means not only studying disease. One of the best-known recent "treasured exceptions" to receive considerable attention is work led by Dr. Helen Hobbs and Dr. Jonathan Cohen of the University of Texas Southwestern, who showed that rare loss-of-function mutations in the PCSK9 gene lead to unusually low levels of cholesterol and protection against cardiovascular disease. Indeed, this focus on exceptions has been at the heart of medical and human genetics ever since the time of Garrod. As we will see at the end of this chapter, however, after a century of focused effort largely inspired by this perspective, human genomics is also now beginning to move beyond it.

Emphasizing the challenge of moving from a recognized genetic condition to the identification of the responsible gene, it would be another century before the gene responsible for the darkened urine that Garrod described was identified. In Garrod's time, and for the many decades that followed, the scientific tools needed to move from the observation of Mendelian inheritance in families to the underlying gene did not exist. A few genes were identified using very laborious approaches focused on faulty proteins. But the work was slow and difficult and in a great many cases simply not possible. That all was to change following Botstein's road map for the systematic identification of disease-causing genes.

As noted, the basis for Botstein's proposal had been known since the early days of the genetics work with fruit flies described in chapter 2. Ever since this early work, it had been clear that there was a theoretical possibility of determining where in the genome a disease-causing mutation is located by studying co-inheritance between disease and allelic variants in the genome. Allelic variants that reside near the disease-causing variant will tend to be co-inherited with disease. But the human genome is very big, with many different groups of linked genes. For anything similar to work for systematic gene mapping in humans, it would be necessary to have genetic signposts distributed widely throughout the human genome, representing all the groups of tightly linked genes near another on each of the twenty-three pairs of human chromosomes. It would also be necessary for it to be possible to readily assay those signposts to determine which forms were present.

In 1980 Botstein and his colleagues described how this could be achieved using naturally occurring human genetic variations, or polymorphisms. Their paper is such a tour de force of clarity and vision that it inspired the career choices of many young geneticists, including me. By determining which of these naturally occurring polymorphisms, spread throughout the human genome, were inherited together with disease status, it would be possible to determine where in the human genome the disease-causing mutation was to be found. Only a few years later the concept was validated when the gene for Huntington's disease was localized by "linkage mapping." The first complete maps of genetic markers began appearing in the 1980s and rapidly evolved in subsequent years to be densely distributed throughout the entire human genome, an early critical output of the Human Genome Project. Before the advent of linkage mapping in humans – or positional cloning, as it is often called (reflecting the use of only knowledge of the position of genetic markers to find disease-relevant genes) – not more than about a hundred human genes had been tied to specific diseases. By the beginning of the era of high throughput sequencing of human genomes twenty years later, that number had grown to around two thousand.

A decisive landmark was reached in 1989, when the gene for cystic fibrosis was identified by a large group of collaborating investigators, including Dr. Francis Collins, who would go on to lead the Human Genome Project. The identification of cystic fibrosis transmembrane conductance regulator (CFTR) as the gene carrying mutations responsible for cystic fibrosis was a key step in human genetics, as one of the earliest discoveries regarding an important (in terms of number of patients affected) and devastating genetic disease, using the road map outlined by David Botstein. But it is also a landmark in illustrating the challenges of moving from identified genetic causes to treatments that are informed by those genetic causes. It was not until 2012 that the first treatment targeting the underlying cause of disease was introduced for cystic fibrosis, a generation after the gene was discovered to carry disease-causing mutations. Although the end of the twentieth century witnessed an explosion of gene discovery for Mendelian diseases, the development of mechanistically targeted treatments has lagged dramatically far behind. There are many reasons for this, but one is unfortunately economic. Many of those diseases affect a very small number of patients and are therefore not attractive candidates for drug development efforts for many companies, although this is currently changing for a number of reasons.

—

One important class of genetic disease, however, is entirely refractory to the positional cloning strategy envisioned by Dr. Botstein and his colleagues. Recessive diseases, no matter how devastating, will often show multiple affected individuals in families, because carriers of the mutations are healthy and can have children. That means it is possible to look at large families to correlate disease status with markers distributed throughout the genome to pinpoint the location in the genome of the gene responsible for a disease. Brand-new mutations that cause a disease severe enough to compromise reproduction, however, are completely refractory to this design. Family members almost always have only a single affected individual, and there is no possibility of using the linkage-mapping strategy outlined by David Botstein and his colleagues. The identification of dominant disease genes carrying brand-new, or de novo, mutations had to await the deployment of next-generation sequencing approaches, which I will describe after a brief detour into common genetic variation and its role in diseases more common than Mendelian ones.

GWAS and the Genetics of Common Disease

Although remarkable progress was made in determining the causes of Mendelian diseases after the work of Dr. Botstein and the development of dense genetic maps in humans, understanding more common and complex diseases proved much more challenging. Many geneticists have long held that there is a sharp distinction in terms of the underlying genetics among diseases that tend to follow Mendelian patterns of inheritance, like the ones we discussed above, and ones that often do not follow such patterns, which include most common diseases. In fact, we know now that the divide is much less sharp than was imagined for many common diseases, as addressed in the final section of this chapter. But it is nevertheless clear that for many common diseases the influences on who does and does not have disease are much more complicated than a mutation in a single gene. The first statistically principled and systematic effort to study the basis of the more common diseases, like depression, schizophrenia, heart disease, type 2 diabetes, epilepsy, and many others, focused on the common variation in the human genome. Genome-wide association studies (GWAS) represented the first attempt to relate a whole category of genetic variation to the risk of these more common and presumably more complex diseases.

At their core, GWAS are predicated on the idea that common genetic diseases are influenced, at least in part, by common genetic variants. When scientists began to map out how to perform GWAS, there was, to many, good reason to believe that common variants might play an important role in the risk of common diseases. First, there was empirical evidence supporting the proposition that common variants can have important effects on the risk of common diseases. A seminal contribution was made by the American geneticist and neurologist (and inveterate iconoclast) Dr. Allen Roses. In the early 1990s, Dr. Roses led a series of investigations demonstrating that a common variant at the ApoE gene increases the risk of late-onset Alzheimer's disease manyfold. ApoE encodes a particular type of lipoprotein, one of whose functions is to combine with fats and help carry them through the blood. How variation in this gene strongly influences the risk of Alzheimer's disease remains poorly understood. At the time leading up to GWAS, many geneticists, including me, took the view that the effects of common variants in this gene, which has such a strong effect on a common disease, would surely *not* be the only such effects to be found in the human genome.

87

Theoretical arguments could also be marshaled in support of the important role of more common variants in the risk of common diseases. One category of argument is a version of the balanced hypothesis of Theodosius Dobzhansky, as described in the previous chapter. As an example, a sometimes popular theory during the run-up to GWAS was the concept of the thrifty gene, proposed by the geneticist Dr. James Neel to explain varying risk of diabetes in different population groups. The idea is that in low-nutrient environments selection would favor energetically economical alleles that would conserve the few calories people were able to find. But in high-nutrient environments, with the Golden Arches of McDonald's beckoning on nearly every block, these energetically economical alleles would be a significant disadvantage, leading to obesity and the many associated diseases, in particular diabetes.[2]

A final theoretical argument was marshaled by geneticists such as David Reich and Eric Lander, who argued that the specific demographic history of humans could lead to relatively common alleles with pronounced effects on the risk of disease. This argument was contested by other leading analysts, who arrived at an essentially opposite conclusion, illustrating that the predictions of our models remain highly sensitive to the assumptions that are made. Altogether, however, the stage was set for a comprehensive investigation of how common variation influences the risk of disease.

A major breakthrough in terms of implementing such a program was provided in 2001, when an exceptionally talented young research associate, Mark Daly, in the Eric Lander group, noticed a pattern of human genetic variation that had until then entirely escaped attention, illustrating again the importance of paying attention to the younger members of research groups. It had long been known that variants near each other in the human genome tended to be associated with one another, a phenomenon geneticists call linkage disequilibrium, which is reflective of the linkage groups first uncovered by Sturtevant. In fact, linkage and linkage disequilibrium are related but quite distinct concepts. Moreover, the term "linkage disequilibrium" is something of a misnomer that continues to trip up students in genetics courses. Linkage has to do with the tendency to be inherited together, with the tendency controlled by the rate of recombination between the sites of variants, which Sturtevant used to work out the first linkage maps for the fruit fly. Linkage disequilibrium, however, relates to the tendency of variants at different places in the genome to be found together in a population (as opposed to the progeny in a cross) more often than expected by chance. The relationship stems from the fact that when variants are only loosely linked with each other (that is, not near one another on chromosomes), the association in populations is rapidly broken down by recombination among variants.

There are many reasons that linkage disequilibrium exists in the human population, but the salient point for our purposes is that when there is strong linkage disequilibrium, what is present at one site in the genome can tell you what is present at another site. Imagine that one site in the genome has alleles A and a and another has alleles B and b. If there is strong linkage disequilibrium, knowing the genotype for an individual at the A locus tells you with confidence the genotype at the B locus. Until Mark Daly's work, geneticists tended to evaluate the *average* drop-off of this association as a function of the physical distance in the genome among the sites considered. That is, analyses would consider *all* sites in the genome at the indicated distance together and average the associations observed for variants at different specified physical distances. Mark Daly, in a dramatic illustration of the importance of looking closely

—

at your data, noticed that within *particular* genomic regions the drop-off of linkage disequilibrium was rarely gradual. Instead, there were often sets of variants that all showed strong association with one another over relatively long stretches. But then, suddenly, the strong association would drop off dramatically, not gradually. The variants that were associated with one another came to be referred to as islands or blocks of linkage disequilibrium. Daly's observations had a galvanizing effect on the field and helped to create enthusiasm for the development of a kind of common variant map, where each of the blocks of associated variants was to be marked off by a sentinel variant (which came to be called a tagging single nucleotide variant, since the sentinel variant gave information about, or tagged, its associated partner variants). To test whether any variant in the group is associated, it would be possible to test just the sentinel tagging variants. This resulted in a far more economical study design than would have been possible without leveraging linkage disequilibrium in this way. And so was born GWAS.

In the early days, these studies were treated with exceptional enthusiasm by most (though not all) geneticists. For the first time, statistically reliable associations were derived between common variants and many complex human traits. Of nearly equal importance, the development of an appropriate statistical framework for relating common variation to variation among people rapidly revealed that most of the earlier associations reported between specific candidate genes and complex traits were incorrect. The reason this happened, in retrospect, is obvious.

There are about ten million common variants in the human genome. That is, there are about ten million sites in the human genome, out of the around three billion sites, where the rarer form in the human population is present at least around 5 percent of the

—

time. Furthermore, among these ten million variable sites, on average about ten variants are associated with one another. Because of this pattern of association among variants, a single sentinel variant represents, on average, ten other common variants. This means that if a single study evaluates the role of common variation in, say, the risk of depression, the human genome affords about one million different independent tests of the relationship between common variation and depression. Each one of those tests represents an opportunity for association.

In the era of work focused on candidate genes, different scientists in different places would test their favorite genes and, of course, they would tend to publish results when they appeared to be significant, considering only the genes and the variants in them that they liked as candidates. Meanwhile, other scientists tested the genes and variants they liked best. But this way of working affords no way to appropriately track the number of comparisons that were really being made by the community as a whole. Researchers might report, for example, that one or another polymorphism in some gene that is important in the brain showed a moderate association with depression, schizophrenia, or some other condition, not correcting for all the other comparisons that other investigators were performing on other genes that they favored. Remarkably, when the first GWAS appeared for many of these diseases, the first thing they showed was that when statistics were properly corrected for all the possible comparisons in the human genome, nearly the entire set of earlier candidate gene associations were shown to have no real association for common variation with the different common diseases. For some conditions, including depression and schizophrenia, the entire set of earlier candidate genes was flatline. This finding allowed researchers to stop working on genes that were not truly associated

—

with the conditions, although in some cases, rather grudgingly. In a very sobering illustration of how science works, large bodies of the work of some scientists were shown to rest on statistical foundations that were not just shaky, but quite simply absent.

In addition to invalidating the vast majority of the earlier candidate gene studies, GWAS also rapidly identified a number of new and statistically reliable associations with most common diseases. These have indisputably stood the test of time and represent real associations. In the early days of GWAS, these findings were considered critical to developing new biological insights into these diseases that would inform drug development for the diseases. Indeed, many prominent genomicists aggressively pushed these findings on sometimes skeptical drug developers in the pharmaceutical industry. But relatively rapidly the new findings revealed a new problem. For most common diseases, the effect sizes of the implicated variants were very modest, nothing like the ApoE effect that Dr. Roses had described. This, by itself, is not an insurmountable challenge. After all, a common variant with a modest effect on the activity of a gene might yet identify a target for drug development that, when modulated more strongly pharmacologically, could realize a greater effect on the symptoms of disease than is reflected in the risk associated with the common variant. For most traits, however, the effects of common variants were so small that it soon became apparent that an inordinate number of common variants would be required to explain the heritability of most common conditions. Writing during the early days of GWAS, I highlighted this problem with a simple mathematical extrapolation of the effect sizes observed for type 2 diabetes and height to estimate how many common variants would be required to explain the genetic component (or heritability) of those traits.[3]

To make this concrete, consider the early output of GWAS for height. The most important common variant identified is responsible for a third of 1 percent of the variation of height in the general population, and the remaining associated variants all impart smaller effects, mostly very much smaller. By simply extrapolating those effect sizes, I found that 93,000 common variants would be required to explain the heritability of height. I concluded, very controversially at the time, that the effect sizes of common variants were not only small—they were *too* small. And the arguments about realizing a greater effect through pharmacological manipulation could not hope to overcome this fundamental problem. If GWAS ultimately identified such a large number of associated variants, spread throughout the genome, the genetic studies of common variation would point at nearly everything in the human genome. And, I suggested, in pointing at everything, genetics would in fact point at nothing. What I meant by this was that in terms of leads for drug development and in terms of insights into disease pathophysiology, all the genetic studies would be telling is that the answer is in the genome. Hardly an actionable result for companies trying to decide which genes in the genome encode the *best* targets to investigate for treating common diseases and are worthy of the hundreds of millions of dollars often needed to develop and test candidate treatments. A decade later, using considerably more data and more sophisticated analyses, this conclusion about GWAS was both confirmed and extended by Dr. Jonathan Pritchard and his colleagues, who argued that these studies were in the process of implicating all genes expressed in tissues relevant to each of the diseases under study.[4]

Today, it is becoming increasingly accepted that most of the associations derived by GWAS are very difficult to translate into novel biological insights into the causes of diseases or into novel directions for therapeutic intervention. Despite this disappointment, increasingly now accepted as reality, the data from GWAS have ended up providing an unanticipated application. While generally proving of limited value in terms of disease pathophysiology and also in terms of guiding the development of new therapies, they have provided important new data in predicting the risk of disease. As just one example, those with the highest burden of common variants increasing the risk of coronary artery disease have been estimated to have three times the probability of coronary artery disease compared with the general population, which is an effect similar to that of some rare single mutations causing hypercholesterolemia. Similar results have been derived for a broad range of common diseases, indicating that the complement of common variants in individual carriers, in totality, provides important information alongside traditional risk factors and family history and may well constitute a target for reproductive genome design in the future, as discussed in the next chapter. These risk scores, based on the complement of common variation an individual carries, are now called polygenic risk scores, and I address their relevance in reproductive genome design in the final chapter.

Infinite Variety: Sequencing, Genomic Individuality, and Disease

The most significant, and probably, in some ways, final, technological advance in the genetics of human (inherited) disease has been the ability to sequence nearly complete human genomes. Now, instead of indirectly searching for the causes of disease by looking at markers of genomic variation, next-generation sequencing has made it possible to find nearly every variant present in the genome of any

—

person, economically and quickly. Two key advances have facilitated the routine sequencing of human genomes. The first, of course, is the output of the Human Genome Project. The fundamental aim of the project was straightforward — determining what sequence is present at all of the approximately three billion sites in the human genome. Of course, you are aware by now that there is no single human genome. Instead, every human has his or her own genetic makeup. But our genomes are very similar to one another. In fact, if you choose a place in the genome at random, the odds are very good that any two people selected will have the same thing present at that random place in the genome. Although the aim of the Human Genome Project was simple, and eventually of clearly proven value, the implementation was daunting. The case for a systematic effort to determine the exact sequence of the human genome was considered by a panel commissioned by the National Research Council.[5] Their report was published in 1988 and outlined the benefits of determining the sequence of the human genome. They concluded with remarkable restraint that "a special effort in the next two decades will greatly enhance progress in human biology and medicine."

Although the technical challenges were considerable, and too detailed to recount here, one of the most important challenges was more sociological than scientific. Until the Human Genome Project, biological research had been largely focused on individual laboratories. While reagent- and data-sharing were the norm, funding and, critically, project design were done almost entirely in labs. In this model, individual labs, driven by the vision of individual scientists, made the key decisions that cumulatively advanced science. The Human Genome Project, however, with a very well-defined goal of determining the sequence of the human genome, would have to be a "big science" effort, drawing in thousands of investigators dedicated to a single clearly defined goal. This was largely a new way of working in biology and led to many concerns about the suppression of the creativity of individual groups pursuing their own interests. The sometimes acrimonious debate about the trade-offs between "team science" and investigator-led science continues to this day.

Eventually the first draft of the map of the human genome was announced by President Bill Clinton and Prime Minister Tony Blair in 2001. To get an idea of just how much information was represented at the completion of this project, if you were to sit down and read out your own complete genome sequence, base by base, it would take you an entire month, if you read at about the same speed I do, anyway. The map of the human genome provided by the Human Genome Project provided a reference (constantly updated) against which all current sequencing data are compared. Given that nearly all biomedical researchers now working in human genetics and human biology make continuous use of the map of the human genome, very few contest the value of the Human Genome Project. Many questions remain, however, about what kinds of questions are best addressed by the big science approach pioneered by the Human Genome Project in biology, and what kinds of questions are best addressed by individual labs going their own ways.

The second critical advance was next-generation sequencing (NGS). The way sequences were determined in the Human Genome Project was slow, laborious, and expensive. Although the overall cost of the project was about $3 billion (US), amounting to about a dollar per base, the direct sequencing costs were something like $1 billion for a single complete sequence (in fact, a mosaic sequence based on a handful of individuals). Today, a human genome can be sequenced in hours for under a thousand dollars, and the small pro-

---

Nucleotide Sequence (240 nt):

ATGTCTTCCCAGCAGCAGCAGCGGCAGCAGCAGCAGTGCCC
ACCCCAGAGGGCCCAGCAGCAGCAAGTGAAGCAGCCTTGTC
AGCCACCCCCTGTTAAATGTCAAGAGACATGTGCACCCAAAA
CCAAGGATCCATGTGCTCCCCAGGTCAAGAAGCAATGCCCAC
CGAAAGGCACCATCATTCCAGCCCAGCAGAAGTGTCCCTCAG
CCCAGCAAGCCTCCAAGAGCAAACAGAAGTAA

Translation (79 aa):

MSSQQQQRQQQQCPPQRAQQQQVKQPCQPPPVKCQETCAPKT
KDPCAPQVKKQCPPKGTIIPAQQKCPSAQQASKSKQK

Figure 4.1. Sequence of the SPRR4 gene and protein. (Drawn by David B. Goldstein)

portion of the genome that encodes protein can be sequenced for a few hundred dollars.

By now you probably have often read about the sequencing of human genomes but may not have a precise picture of what that really means. In fact, it is simpler than you probably imagine. To illustrate exactly what we get sequencing a human genome, the reference sequence for SPRR4, one of the smallest genes in the human genome, is shown here, followed by the sequence of amino acids that this gene encodes (figure 4.1). In fact, if you returned to the diagram of the genetic code we saw in chapter 2 (figure 2.3), you would be able to determine for yourself the amino acids that this gene encodes. To get you started, recall that the overwhelmingly predominant starting nucleotide word, or codon, is ATG, which you will see right at the beginning of the sequence (only the protein-coding part of the gene is shown). In the genetic code table you will

---

find that this codon encodes the amino acid methionine, which is indeed the first amino acid indicated in the protein (denoted by M). If you work through the rest of the codons, you will be able to fully reconstruct the amino acid sequence of this little gene.

Of course, we do not do this kind of work by hand. Even for a very small gene, it is a laborious task, which I suspect you have not bothered to complete (I certainly have not by hand). And if I were to have chosen to represent the largest gene in the human genome, Titin, instead of one of the smallest, the protein-coding portion of the gene would stretch over more than thirty pages, which would certainly lead to objections from my publisher. Instead, when we sequence human genomes, computer programs do this work for us. How this works in practice today is that computer programs first determine what part of the reference genome is being considered, in a process called "alignment," and then programs determine all the differences a test sample carries in the aligned region compared to the reference. The reference is, essentially, the sequence that was first determined by the Human Genome Project.

This technological advance in sequencing has made it possible to start with a child with a serious disease and to scour his or her genome for significant mutations, and then often to identify the clear genetic cause of the child's disease. One early important application of NGS was to fill in our knowledge of disease-causing genes that could not be found by the linkage studies that had proved so effective for recessive diseases. As noted, when de novo mutations are responsible for diseases that compromise reproduction, families usually have only a single affected individual. A rare exception to this can sometimes occur when the responsible mutation is mosaic in one of the parents and thereby transmitted to more than one offspring. Either way, for these kinds of mutations, the analysis of the co-inheritance of genetic markers and disease in families does not work to track down the disease-causing mutation. But these genes with disease-causing de novo mutations are readily identified by sequencing the genomes of affected individuals that are unrelated to one another and by searching for disruptive mutations that occur in the same gene in the affected individuals but not in unaffected individuals. We simply generate a sequence like that shown on page 97 and tally up the differences to the reference we see in all affected individuals in all genes. The gene that has a clear excess of important mutations in all or most of the affected individuals is the gene carrying disease-causing mutations.

As an illustration of how this works, about ten years ago we started a study of an unusual childhood disease called alternating hemiplegia of childhood (AHC), which involves a range of neurological conditions often including seizures and the characteristic feature of paralysis of only one side of the body that alternates between sides at different times. By sequencing the protein-coding part of the genomes of only ten patients, we found that seven of them carried de novo mutations that changed the protein sequence of a single gene, ATP1A3. Because such mutations are rare in the human genome, we knew we had found the gene. Similar approaches rapidly identified many hundreds of new disease genes, many due to de novo mutations that could not be detected through linkage and some recessive genes that were too rare to have been uncovered during the linkage era, like the NGLY1 gene responsible for Bertrand Might's condition, caused by an inherited genetic mutation.

Next-generation sequencing not only allowed the discovery of many new disease genes but has also allowed diagnostic sequencing to be deployed even without any knowledge of the gene likely affecting the individual patient. Until NGS came along, a patient

—

might have one or a handful of genes sequenced if the clinician suspected the genetic cause. But, as we saw earlier, in many cases of presumed genetic diseases, there is no way to guess the relevant gene. Today this is no longer a barrier, and increasingly either whole-exome or whole-genome sequencing is used to search for mutations wherever in the genome they might be, whenever a strongly genetic condition is suspected.

This approach has taught us that genetic diseases often present in unexpected ways. In chapter 1 there was the example of Cara Greene, who had a rare genetic disease that could not be recognized clinically for what it is. This is an increasingly common observation for rare pediatric diseases and has led to a growing recognition of the need to sequence the genes of all children with suspected genetic conditions. But the impact went beyond rare childhood diseases. In fact, a surprisingly high proportion of adults with complex diseases turn out to have unrecognized Mendelian diseases. One of the most striking examples is chronic kidney disease, an archetypal example of a common complex disease affecting as many as one in every ten adults today. In a recent study led by my colleague Dr. Ali Gharavi at Columbia University, we showed that 10 percent of adults treated for chronic kidney disease at the medical center managed by New York–Presbyterian Hospital and Columbia University have disease because of mutations in just one of many different genes that when mutated can lead to kidney disease. Similar examples have now been reported in diseases ranging from epilepsy and autism to heart failure.

The other major innovation facilitated by NGS is the development of the science of population genomics. Before the development of NGS, we had no way to assess the pattern of genetic variation throughout the human genome in large numbers of people or to systematically relate that variation to the differences among people.

However, the development of large databases of people whose genes have been sequenced, usually as part of research studies, has permitted the development of entirely new tools for interpreting human variation. As one key example, these data sets have allowed us to learn what parts of the genome can carry mutations without adverse effects and what parts of the genome, when mutated, cause harmful effects that are kept out of the population by natural selection. In 2013, Dr. Andrew Allen, Dr. Slave Petrovski, and I introduced a framework for identifying places in the human genome that are under such strong selection that functional variants cannot become common in the human population. We called those regions that are depleted of functional variation relative to expectation "intolerant" and showed that disease-causing mutations are enriched in such intolerant regions. The basic idea is simple (as outlined in figure 4.2). You look at thousands of people who have had their genes sequenced and count up the number of variants of different types throughout the genome, then compare those variants to how much variation is expected. Regions of the genome with less variation than expected are where disease-causing mutations are most likely to be found, and this scoring system has proved invaluable to interpreting the genomes of patients.

Most recently, Kate Stanley, then working in my lab at Columbia University, led a study that used this framework to identify mutations responsible for stillbirth. These mutations would have been very difficult to find otherwise because many are found in genes that are not known to cause any post-natal disease, presumably because these mutations, when present, do not permit development of the fetus to term. But, knowing which genes are completely intolerant

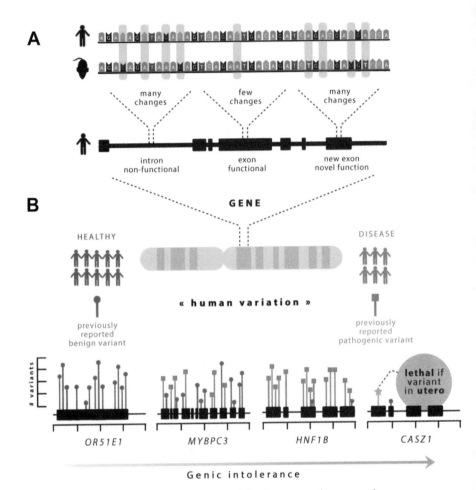

Figure 4.2. Schematic representation of the divergence of sequences between species and the pattern of variation within the human population, showing genes that are tolerant of genetic variation to the left and those that are intolerant to the right. (Modified from Kate E. Stanley et al., "Causal Genetic Variants in Stillbirth," *New England Journal of Medicine* 383, no. 12 [September 2020]: 11071–116. www.nejm.org/doi/10.1056/NEJMoa1908753. Copyright © 2020 Massachusetts Medical Society. All rights reserved.)

to extreme mutations, we were readily able to show an excess of such mutations in precisely those genes. This work showed that there are many genes relevant to stillbirth that are yet to be discovered that will not be uncovered through the study of post-natal disease. Despite how common miscarriages are, genetic analyses lag far behind other important presentations, perhaps due to the continuing discomfort of affected families in talking about their experiences. Not long ago, at the end of 2020, Meghan, the duchess of Sussex, wrote in the *New York Times* about her own experience of a miscarriage. Hopefully her poignant and articulate description of the pain, and even shame, that so many families have felt will help make miscarriage something that is more acceptable to discuss, study, understand, and help to prevent.

The other major advance that NGS facilitates is the development of the data necessary to systematically relate genetic variation to the differences among people. A vision for a program of research to do this was outlined by another committee of the National Research Council (NRC) at a meeting held twenty years after the Mapping and Sequencing committee report came out. The report of the recent committee is titled *Toward Precision Medicine: Building a Knowledge Network for Biomedical Research and a New Taxonomy of Disease,* and among other things it helped to launch the American precision medicine initiative now known as the All of Us Research Program, and also a transition in nomenclature from personalized medicine to precision medicine.[6] Perhaps the best summary of this second report is provided by Maynard Olson, the only individual who participated on both committees, who has outlined the key conclusions of both committees for those interested in a more detailed description.[7] Olson emphasized that the key recommendation of the more recent committee, on which I also served, was the

—

proposed generation of an information commons that would allow researchers to access both genome data and data about health outcomes on a population scale. We argued that only by accessing the information available on people through their regular interactions with the delivery of health care could we develop a science that would systematically relate genetic variation to the variation among people. At the time of this writing, this population-based approach to genomics is rapidly advancing, with large-scale initiatives in multiple countries seeking to make paired genomic and health-care data available to researchers. As described in the next chapter, it is this information, developed on a truly global scale, that we most need to inform decisions about reproductive genome design.

Chapter 5

WRITING THE GENOMES OF OUR CHILDREN

It is now clear that we are on a path to being able to reliably identify, in the genomes of people, the primary causes of many of the strongly genetic human diseases. Genes have now been found that carry mutations causing both neurodevelopmental diseases that strike early in life, such as autism, epilepsy, developmental delay, and intellectual impairment, as well as important subsets of later-onset diseases including many cancers, forms of cardiovascular disease, neurodegenerative diseases like Parkinson's disease and ALS, and kidney diseases.

What is, unfortunately, far less clear is how many of these diseases we will be able to effectively treat once the underlying causes have been identified. Despite a number of important examples in which identifying the precise cause of disease tells us directly how to treat that disease in an individual patient, it already appears clear that the majority of diseases that are tracked to underlying genetic causes will not be completely rectified by any conceivable pharmacological or other therapeutic intervention after birth. To the extent that this central thesis holds true, that it will ultimately prove much

—

easier to determine the genetic causes of disease than to effectively and fully treat them, the focus of intervention, at least for strongly genetic diseases, must eventually shift from treating patients that carry those mutations to identifying those causes of disease and ensuring that our children do not have them.

This claim, while perhaps not universally popular, as I suggested earlier, is frankly self-evident. There will clearly be a very substantial market among prospective parents for any tools that can ensure the non-transmission of variants causing diseases that cannot be comprehensively cured. This is not to say that it will be easy or that there will not be deeply challenging questions to address, including the fundamental question of what should be considered a disease appropriate for prevention through reproductive genome design. Nor is it to say that we should in any way relax our efforts to treat the genetic diseases that hundreds of millions of people currently endure throughout the world. Those people and their families deserve our continued committed efforts to find effective treatments, as do the many children yet to be born, for many years to come, who will still carry the mutations that cause serious genetic diseases. This is the effort to which my own work life is entirely dedicated, and it is my view that we need much more work on treating these diseases, not less, because there will be some distant future in which it will not be needed. But that distant future will surely come — in fits and starts and with plenty of missteps along the way — but it will come. The aim of this chapter is to convince you of this and to emphasize some of the challenges we must address as we rush — or, far preferably, walk deliberately — toward that future and what I am calling the end of genetics.

This reality was captured poignantly by Matt Wilsey, the father of Grace Wilsey, one of the first children identified with NGLY1 deficiency and the one to offer the affirmation of hope to Bertrand Might in his last hours of life. When asked about the potential of CRISPR (clustered regularly interspaced short palindromic repeats) to fix the genomes of children before they are born, he was unambiguous in his answer. In *Nature* in 2016, Erika Check Hayden reported that Matt "says that if he had had the chance to detect and fix the mutation in Grace's genome before she was born, he would have." Note that nowhere are there expressed any concerns about the effect of such editing on Grace's identity. This is about the elimination of a devastating disease that disrupts who Grace is otherwise. Matt also wisely notes the frustration that the "debate over editing embryos seems to have monopolized discussions about the technology." He goes on to say that he is hopeful that gene therapy might help Grace in the years ahead, not only future generations. And this is a goal we must always keep foremost in mind, even as we prepare for a new world in which such mutations will be eventually excluded from human populations.

With that call to arms to continue and indeed expand our efforts to treat genetic diseases, I now turn finally to the evolving technologies that will allow us, eventually, to ensure that genetic variants that are judged undesirable are never transmitted to children. And I also consider exactly what variants might qualify as undesirable, both those determined with high levels of confidence to cause serious diseases, and also those judged only as uncertain in their effects, perhaps the most vexing category to consider. It is now clear that this technological ability is within reach. It has moved progressively from the fantasy of Aldous Huxley to the science fiction of the film *GATTACA*, to being within the clearly predictable future. And it is also clear that a great many parents throughout the world will want to make use of these new technologies to ensure that their children do not carry any mutations that are guaranteed to cause serious genetic diseases. I consider it likely that fewer, but still many, parents will also choose not to transmit variants that are in a more ambiguous category, as I will describe. My focus now is on laying out how these technologies work and how they might be deployed. Once this framework for these technologies is established, we can address the key challenge of deciding what should be eliminated from the human gene pool, what should be introduced or adjusted, and what we are best advised to leave to chance and circumstance. My central thesis is that these questions represent one of the most fraught and complicated challenges that humanity has ever faced. We will soon be able to prevent the most serious genetic diseases from appearing in our children. For anyone affected by serious genetic disease, this can be viewed only as one of the most important advances in the history of medicine. After all, there is nothing that parents want more than a healthy child, and parents will appropriately do all that they can to ensure that their children are healthy and remain so.

At the same time, we should all recognize that the last thing anyone wants is to create genetically homogeneous children scrubbed of genomic individuality. Our quirks of individuality are in part environmental, to be sure. But they are also in part genetic. And these quirks of individuality are part of the fabric that unites families. They are there in the magic of seeing a beloved grandparent in the arching eyebrows of a grandchild's smile. When we manipulate this genomic individuality, we risk not only unforeseen medical consequences but also unforeseen emotional consequences. Deciding the balance between the preservation of genomic individuality, respecting the sometimes unexpectedly beautiful quirks of genomic fate, and ensuring, as best we can, healthy children is perhaps the central challenge humanity will face in the years ahead. But there can be no careful assessment of what *should* happen without knowing what *can* happen. We must address not only the genetic changes that will be targeted but also the technologies that will be used to engineer the genomes of our children at will. Above all, we must make clear the singular challenge we face in getting the balance right between reducing the burden of genetic disease and preserving the bountiful and beautiful genomic diversity of our species. Above all, I part company with some of my colleagues in believing that it is our responsibility as professional geneticists to make clear that reproductive genome design is on the way. I also part company with some professional geneticists in believing, emphatically, that it will not be up to geneticists at elite universities in the Western world to decide what will be done. It will be up to parents everywhere.

There are four key elements of this, our final story of how the genomes of our children will be engineered. I begin with examples of the kinds of mutations and genetic variants that we may consider targets in reproductive genome design. Next, in order to ground our discussion in how much is already possible, not depending on any technological advances at all, I outline what can be done with the technology available right now, which would include a most remarkable and nearly comprehensive elimination of at least two key classes of inherited diseases, specifically recessive Mendelian diseases and dominant diseases that are passed on from a parent who carries the mutation. Third, I describe the evolving science that will allow us to go well beyond what is currently possible in reproductive genome design. The aim here is to make clear what is certainly coming at some point and that we must be prepared for. I end the story with the fourth and most speculative part of the book. Here I try to predict what we are likely to edit in the genomes of our chil-

—

dren in the near and medium future and what we need to do in order to avoid potentially tragic experimentation in reproductive genome design. Consistent with my approach throughout this book, I seek to empower readers to formulate informed views by providing an accessible description of the science that underpins the future of reproductive genomic design.

Reproductive Genomic Design: What Are We Targeting?

In the previous chapters I introduced the core of contemporary human genetics, including how we learned the way that genetic information is encoded and how we have slowly developed an understanding of key genetic differences that are responsible for causing serious diseases in a significant proportion of the human population. We have learned about mutations that cause devastating childhood diseases if left untreated and how the majority of those genetic diseases remain untreated today, even when they are correctly diagnosed. We also discussed how, even when there are effective treatments, the challenge will often be to treat the diseases early enough to correct the developmental effects of the disease-causing mutations. This means that, when deployed at any point post-natally, even the most targeted of treatments, including gene therapy, will often not fully correct the effects of the harmful genomic alterations.

Based on what we already know today, we can clearly identify important classes of disease-causing mutations that could be targeted for non-transmission, which I believe would and should be largely uncontroversial. We can also identify classes of variation for alteration that could be pursued, and might be pursued, but would have consequences that are hard to predict.

—

Recessive Genetic Diseases

At the time of this writing, we know of 2,149 genes that cause autosomal recessive diseases out of the approximately 20,000 human genes.[1] As you will recall from the first chapters, these diseases are a result of mutations in genes that generally can tolerate a single mutation in one of the two alleles without causing serious problems. These diseases, therefore, act like Mendel's wrinkled peas. If a plant has at least one copy of the smooth-pea form, its seed shape will be smooth. But if a plant has two copies of the wrinkled variety, the seed shape will be wrinkled. These diseases act like the wrinkled form, in fact exactly like the wrinkled form. This holds for all recessive diseases that are caused by genes found on any chromosomes other than the sex chromosomes; any chromosome that is not a sex chromosome is called an autosome, and thus we call these diseases autosomal recessive diseases. This pattern of inheritance exactly follows the wrinkled form of Mendel's peas, illustrating the generality of his rules. With two recessive mutations, the pea's seed shape is wrinkled. As for these genes in humans, two mutant alleles mean that the individual will express the disease.

We know of a further 288 genes that are on the X chromosome and cause disease, in a pattern of inheritance that we call X-linked. Since males have only one X chromosome and females have two, most affected individuals are males who have inherited a mutation from their mothers. One of the earliest recognitions of the particular inheritance pattern these mutations cause can be found in the Talmud, which provides for an exception to circumcision if babies on the mother's side of the family, but not the father's, suffered catastrophic bleeding upon circumcision. As you will have realized, the rabbis had recognized the disease we know today as hemophilia. It

—

is familiar in particular because Queen Victoria passed a disease-causing mutation on her X chromosome to her daughters and from there to male royals throughout the European continent.

As detailed earlier, virtually all of us carry single mutations in one or more of the approximately two thousand genes that can cause recessive genetic diseases. In populations that are largely outbred, meaning that very few children are born to close relatives — for example, few of the parents that produce the children of the population are themselves first-, second-, or third-degree relatives (such as siblings, aunts and uncles and nephews and nieces, or cousins, respectively) — the burden of such diseases has been estimated to be 1.7 out of every thousand live births for diseases that manifest within the first twenty-five years of life.[2] This may sound like a small proportion out of each one thousand humans, but this class of disease alone translates to a staggering 13 million people worldwide, about 50 percent more people than the population of New York City today. In fact, the global burden is even higher, because some parts of the world have substantial amounts of inbreeding, in which parents who are third-degree relatives or closer have children, and such pairings of relatives greatly elevate the rate of recessive disease. The reason is obvious. In unrelated individuals, the genes in which the parents carry single-copy disease-causing mutations are usually different. In related individuals, these genes are much more likely to be in the same genes (because the same mutation has been inherited from a recent ancestor they have in common).

These diseases include some of the most common genetic diseases that are well known in the general population and that we have described previously, such as cystic fibrosis (affecting about one out of every 2,500 people of European ancestry), sickle cell anemia (affecting about one out of every 600 African American babies born

—

112

today), and Tay-Sachs disease (for which about one out of every 30 Ashkenazi Jews today carries one disease-causing mutation). Also included in this group are much rarer conditions including Brown-Vialetto-Van Laere syndrome, which afflicted Cara Greene until she was accurately diagnosed and treated, and NGLY1 deficiency, a deficiency of N-glycanase 1, an enzyme that normally helps the body remove proteins that are not functioning properly, which Bertrand Might eventually succumbed to at twelve years of age. One simple observation makes clear the impact that reproductive genome design will eventually have in the human population, whatever one's view of how long further technological developments will take. The common recessive diseases we have just named, along with the roughly two thousand other such diseases, could be entirely removed from the human population using technologies that are available today. The elimination of this contribution to the global burden of mortality and morbidity is, in my view, an unambiguously positive first goal for the reproductive genome engineering to come. Exactly how this can be done with existing technologies is something I will describe in the following section after discussing the second category of genetic diseases addressable with technology available today.

Dominant Diseases

Most of the remaining strongly genetic diseases follow the pattern of inheritance Mendel described. Today we know of 1,488 genes that cause disease following a dominant mode of inheritance.[3] Here disease now takes the form of the smooth seeds. Any plant with one smooth allele produces smooth seeds, and any child born with a single disease-causing mutation in one of these genes will have the disease. Well-known examples of dominant genetic diseases include neurofibromatosis, Huntington's disease, Marfan's syndrome, Fragile X disease, and the majority of the genetic causes of more severe forms of autism, intellectual disability, and epilepsy. To consider how disease-causing mutations in dominant genes can be addressed in reproductive genome design, we must consider whether the disease-causing mutations are generally inherited from parents or are generally de novo mutations in the affected individuals. Unfortunately, in many dominant-disease genes, the responsible mutations are de novo. This means, usually, that they occur either in the early stages of the embryonic development of the individual or, more commonly, in the gametes of the parents that gave rise to the affected individual. As I discussed previously, everyone carries brand-new, or de novo, mutations. Normally they are benign or only moderately harmful, but sometimes they are in critical parts of the genome and cause devastating diseases. This class of variation represents a much greater challenge for reproductive genome design than do inherited mutations, because in this case the cause of disease is generally not identifiable in the parents in advance of reproduction.[4]

For this reason, complete protection of the genomes of children against this critical class of disease-causing mutations will require the development of new technologies, as described below. The role of de novo mutations is well illustrated by the SCN1A gene, the loss of one copy of which is the most common genetic cause of epileptic encephalopathy, an often devastating early-onset condition characterized by seizures and often progressive loss of cognitive functioning. The SCN1A gene encodes a sodium channel, which is critical to how neurons propagate information by abrupt changes in their electrical potential. Specifically, neurons "fire" by allowing a rapid influx of positively charged sodium ions, which are allowed into cells by sodium channels that open when a threshold voltage is

—

reached. When a de novo mutation inactivates one copy of the SCN1A gene, the result, as far as we know, is invariably seizures and a host of other cognitive and related deficits. Along with SCN1A, there are scores of other genes that can carry de novo mutations causing epilepsy or other neurodevelopmental conditions, including autism and intellectual impairment. In clinics today, when the genomes of children with serious diseases are sequenced, the causes of the diseases are often found to be de novo mutations in genes like SCN1A.

The burden of dominant disease that presents very early in life is estimated to be similar to that of recessive diseases, but there are many dominant diseases that present later in life, and often carriers do not know that they have the mutation until disease presents, if it presents. For these diseases, the affected people will often have children, and they will often do so without knowing they carry a disease-causing mutation that they might transmit to their children. Sometimes they will later find out that they do, and that their children might, and sometimes they do not find out at all during their lifetimes, since many of these later-onset diseases have highly variable presentations. And these diseases can be surprisingly common. For example, autosomal dominant polycystic kidney disease usually presents in a person's thirties and forties, but the onset and severity vary widely. In most cases, disease results from an inherited (or a de novo) mutation in the PKD1 or PKD2 gene, though a small proportion of cases are the result of mutations in other genes. Remarkably, this genetic disease alone is estimated to affect more than ten million people worldwide.

There is one particularly important class of later-onset genetic diseases that the American College of Medical Genetics (ACMG) has judged critical to report to individuals as incidental findings

—

whenever they are genetically evaluated for any clinical reason. What is meant here by "incidental" relates to the reason the individual underwent genetic evaluation. If, for example, the person's genome was sequenced because of kidney dysfunction, findings that do not relate to the genetic causes of kidney disease but were discovered as a consequence of the generation of genome-wide genetic information, typically now done through either whole-exome or whole-genome sequencing, would be considered incidental.

The ACMG currently recommends consideration of a set of fifty-nine genes that can carry mutations that cause serious and often life-threatening conditions that are, critically, judged to be actionable. That is, not only are the diseases serious but they are diseases for which interventions can be staged to ameliorate risk. This is the reason the ACMG has advised that patients and their care providers always be informed of those genetic findings, even if, as described, the reason for sequencing was unrelated to the condition these genes cause.[5] Genes in this set cause a range of conditions including, prominently, an elevated risk of cancer and various forms of heart disease. Included in the set of fifty-nine are the BRCA1 and BRCA2 genes, which can carry mutations that greatly elevate the risk of breast and ovarian cancer and, in men, prostate cancer. Individuals who carry risk alleles in these genes are candidates for enhanced screening procedures or, in the case of women, for prophylactic surgery to reduce risk, which was famously pursued by the actor Angelina Jolie. In a clear and accurate op-ed in the *New York Times* in 2013, Ms. Jolie revealed that she carries a "faulty gene, BRCA1," that "sharply increases risk of breast and ovarian cancer."[6] This gene is on the ACMG list because there are steps that women may choose to take to reduce the chance of disease onset. Ms. Jolie herself chose to have a prophylactic double mastectomy, which she reports has reduced her risk of breast cancer from 87 percent to 5 percent. Although these kinds of estimates are based on what we know about risk associated with mutations and family history and are not generally as accurate for an individual as the precise percentages might imply, there is no doubt that these interventions result in a dramatic reduction in risk for people with disease-causing mutations in either BRCA1 or BRCA2.

Ms. Jolie's brave disclosures helped to make genetic risk and the choices those risks can lead to more tangible and understandable for a wider community. Considering both the BRCA1 and BRCA2 genes and the fifty-seven other genes that the ACMG recommends be considered whenever a person's genome is evaluated, it has been estimated that as many as two out of every hundred people carry a disease-causing mutation in one of these "actionable" genes. At two out of one hundred, this would mean that there are more than six million people in the United States alone who, if their genomes are sequenced, would be found to carry a mutation in one of these fifty-nine actionable genes. Like Ms. Jolie, if they were to learn about their own genetics, they would have choices to make about their personal health. Less appreciated is the fact that they would also have choices to make about reproduction, too. If they were given the option, would prospective parents carrying such mutations choose not to transmit them to their children? Like the mutations that cause recessive diseases, this class of variation (inherited dominant disease-causing mutations) can be addressed today using currently available technologies. And, as in the case of the recessive disease described above, seeking to ensure non-transmission of such mutations to children seems to me a reasonable choice, fully supported by existing science, and unlikely to lead to unexpected outcomes, at least when practiced without error.

—

If we consider a broader definition of important genetic changes, the proportion of carriers increases dramatically.[7] As one striking example, the ApoE gene carries gene variants that can elevate the risk of late-onset Alzheimer's disease several-fold up to fifteen-fold, depending on whether individuals have one or two copies of the most at-risk allele.[8] The most at-risk allele is also very common and quite variable in its representation among populations, showing, for example, a range of frequency from around 5 to around 30 percent in different European populations.[9] The ApoE gene, however, is not on the ACMG list because there are no known effective strategies for reducing the risk of Alzheimer's disease. But though there are no effective interventions to ameliorate the risk of a carrier, completely effective interventions are possible to prohibit the transmission of that genetic risk to a child for individuals carrying only a single copy of the risk allele. For individuals carrying two copies of the risk allele, current technologies do not allow non-transmission of the risk allele, but future technologies certainly will, as I will shortly describe.

A very different category of genetic risk to consider is the risk associated with common genetic variation. As shown in chapter 4, genome-wide association studies have been performed for nearly all common human diseases and for many non-disease traits such as height, weight, educational achievement, hair and skin color, and sexual orientation. These data have been used to develop polygenic risk scores that can provide, based on the full complement of common variants an individual carries, a quantitative assessment of the risk of specific diseases. Although the risk associated with any individual common variant is very small, the differences among individuals with different burdens of risk variants can be substantial. For example, the distribution of polygenic risk scores in the general pop-

—

ulation for cardiovascular disease (CVD) shows that, at the extremes of people with relatively fewer or relatively more risk alleles, the difference in risk can be similar to that of rare disease-causing mutations. At the most extreme percentiles, the elevated risk is comparable to that attached to rare mutations causing hypercholesterolemia, as mentioned in chapter 4. Consideration of polygenic risk profiles can be used with today's genetic technologies in reproductive genome design, albeit with very modest effects currently, as I describe below.[10] Expected technological advances, however, will allow much more dramatic uses of polygenic risk profiles, though unfortunately with as yet unpredictable consequences.

A final category of genetic variation that parents may wish to consider in their plans for reproductive genome design are genetic variants that influence susceptibility and resistance to infectious diseases. At the time of this writing, this question is of urgent contemporary relevance with the virus responsible for COVID-19 rampaging throughout the world. As I outlined in chapter 4, we know that for many infectious diseases there are gene variants that protect and gene variants that confer susceptibility. Unfortunately, with the exception of diseases that compromise entire arms of the immune system (and that would be covered by the considerations above), other genetic variants are more nuanced in their effects, protecting against some diseases and not others or indeed conferring resistance to some diseases but susceptibility to others. As the COVID-19 pandemic tragically illustrates, however, we have no way of knowing exactly what infectious agent may threaten us and when. SARS-CoV-2 appeared in Wuhan, China, in late 2019. It was presumably transmitted from bats to an intermediate vector yet to be determined and from there to humans in one of the markets selling live animals. Before the outbreak, no one would have predicted that a

—

119

novel coronavirus would soon sweep through the world, choking both intensive care units in scores of countries and global economic activity. Before long, we will subdue this latest particular viral assault on the human population, primarily through global inoculation with effective vaccines. SARS-CoV-2 is not a highly mutable virus, and there is every reason to believe that recently introduced vaccines will prove as effective in deployment as they appear in trials.

But what we cannot know is exactly what will attack the human population next, or how we might defend against whatever it might be. The one thing that we can know about infectious diseases in general is that genetic homogeneity is a bad idea. In the cases of most of the pathogens that have been well studied, some humans are protected. When entire populations can be the most at risk is when those populations have little genetic variation. Humans have substantial genetic variation and do not exactly resemble potatoes in meaningful ways, but the history of cultivated potatoes illustrates the point in caricature form. In the 1800s, Irish farmers began concentrating much of their efforts on one particular potato variety. The "Irish Lumper" was not considered very tasty, but it is easy to grow in lousy soil and has a substantial yield. It rapidly came to play the key part in the diets of people in Ireland, in particular for the poor. Its widespread adoption, unfortunately, also led to a marked reduction in the genetic variation in cultivated potatoes in Ireland, making much of the caloric production for the country vulnerable. In the 1840s, disaster struck in the form of a fungus that wiped out the potato harvest for years, leading to the Great Famine and the loss of a million lives. Had there been more genetic variation among the potatoes cultivated, the national harvest would have been at much lower risk. We cannot say how to protect against any particular pathogen in humans, but we can say that reducing the genetic variation among humans through reproductive genome design could make the species as a whole much more susceptible to particular pathogens. That's something for us to consider in making the choices to come.

I now turn to what changes in the genomes of children, out of the categories considered above, we would be able to reliably make if we employed only technologies already available.

Reproductive Genomic Design: What Can We Do Today?

The story of Tay-Sachs disease illustrates how much the global burden of genetic disease could be reduced using only technologies available today. Mutations responsible for Tay-Sachs are remarkably common among Ashkenazi Jews, with as many as one out of thirty carrying one disease-causing mutation. Before screening for the disease-causing mutations was widely deployed, fifty or more affected Jewish children were born in the United States alone every year, filling specialty clinics in areas with large numbers of Ashkenazi Jews. Today, no more than a handful of affected Jewish children are born each year in the United States and Israel combined. How was this achieved? The starting point is knowing who carries the mutation. In the early days of testing, identifying carriers was based on looking at the activity of the enzyme responsible for Tay-Sachs disease. Today it is based on testing for disease-causing mutations directly. What is done with the information depends on who you are. In some religious groups, this is handled by large-scale testing of the community and a process to assess "compatibility." A well-known example in New York is Dor Yeshorim: Jewish Genetic

Screening. When a community member considers a prospective partner, he or she can provide Dor Yeshorim the codes for him- or herself and his or her partner. Within days word will come back about whether the couple is compatible (meaning that they have no mutations in the same gene) or incompatible (meaning that both individuals have a disease-causing mutation in the same gene). In the case of Dor Yeshorim, these tests are performed before marriage.

In other settings this is handled through the rapidly increasing use of what is called expanded carrier screening, in which the individuals find out exactly what mutations they carry in a set of up to several hundred genes known to cause recessive disease. However couples find out their status, if a couple considering conception is found to have mutations in the same gene, in the case of Tay-Sachs, in the HEXA gene, there are two interventions available. The first is simply to find a new partner, as often happens with the services Dor Yeshorim provides. The second option is to use in vitro fertilization and to select an embryo for implantation that is known to carry at most only one of the disease-causing mutations. How exactly does this second option work?

The story begins with the birth of Louise Brown on July 25, 1978. Before Louise, just about every human alive, and just about each and every human ever born, was conceived the old-fashioned way, which we need not say more about here.[11] Since the birth of Louise Brown, however, a rapidly increasing number of babies have been born following conception not in a woman's fallopian tube en route to the uterus following intercourse but rather in a petri dish. Given how commonplace in vitro fertilization is now, today it is hard to appreciate just how remarkable it was considered to be at the time—and how controversial. In the decades ahead, when we look back on the steps required for systematic reproductive genome

—

design, the birth of Louise Brown may still appear to be the biggest step of all. The technological advances still required, though dramatic, may yet seem modest in comparison to this monumental change in the nature of human reproduction. There really was no way to make a convincing case, then, that in vitro fertilization would in fact work. It was yet another example of exceptional clinical need driving scientific exploration. And this is exactly what will drive the next set of innovations.

Louise's parents, Leslie and John Brown, could not have children naturally because Leslie's fallopian tubes were blocked. But they wanted children. The basic process of in vitro fertilization is to remove an egg from a woman's ovary and combine this with sperm from the prospective father in a laboratory dish, thus permitting fertilization to occur. After monitoring the development of fertilized eggs for several days, one or more pre-embryos are then transferred to the prospective mother's uterus for development into embryos and leading, hopefully, to pregnancy. The reason I highlight this as what may be seen as the biggest leap into systematic reproductive genome design is that we had no way to know if it would work. In fact, I would imagine that most people would have guessed that it could not work, or at least that it would not work well.

When an egg and a sperm cell unite, there is a feat of remarkable developmental magic that takes place. Cells in the human body have mostly the same DNA sequences. But cells look and act dramatically differently from one another. This feat of genomic versatility is achieved by different cells' carefully controlling the expression of the genes that are all present, but not all equally active, through a combination of changes to the DNA itself and to the proteins that ensconce DNA. In the case of the egg, these packaging proteins are histones, as in almost all other cells. In the case of sperm, the role of

—

histones is largely replaced by protamines. These packaging proteins and their characteristics, plus various changes to DNA itself, constitute what we call the epigenetics of the cell. These epigenetic characteristics work alongside a constellation of proteins present in cells, such as transcription factors, to collectively determine the pattern of gene expression in the cell and thereby its identity and function. Sperm and egg cells are both highly specialized cells loaded up with epigenetic changes to help them be the cells they are supposed to be. But when they unite to become a zygote, they need to forget all of that and go back to being essentially just DNA that can turn into anything. How could we know that this would happen as it should outside the body and the very complex environment present there? We could not. And how could we know that there aren't important selection pressures that determine which particular sperm cells achieved fertilization to help ensure that the "right" sperm cells, out of the millions competing, were the ones that would win the competition to fertilize? Again, we could not.

But after forty years of experience, we know that in vitro fertilization does work. As one would expect, there are, indeed, some modest differences in the risks of diseases and other characteristics among naturally conceived children and those born of in vitro fertilization. But we know that more than eight million people alive today were born because of this technological advance, and we know that those people are generally about as healthy and happy as are people born following natural conception.

What is critical for our purposes is what in vitro fertilization (IVF) allows in terms of the elimination of genetic disease. When couples reproduce through IVF, it is possible to genetically test the resulting pre-embryos in a process now referred to as pre-implantation genetic testing (PGT). Various procedures are used. One ap-

—

proach is to allow the pre-embryo to develop to the eight-cell stage and then remove one of the eight cells from each candidate pre-embryo. The removal of one cell at this stage does not adversely affect the pre-embryo's development, and the DNA content of those selected cells is then evaluated to determine the genomes present in the zygotes from which they were derived.

This procedure as of now has two important limitations. First, all that current technologies allow is the selection of the preferred pre-embryo out of the small number that can be produced (a handful at most per round of treatment, limited by the availability of eggs). Second, genetic analysis technologies are still not accurate enough for us to precisely determine the genome of a single cell or just a few cells; we do not get enough DNA for that. PGT currently is used either as a screening procedure to ensure a complete complement of chromosomes in the selected pre-embryo or embryos, or as a diagnostic procedure to make sure that mutations known to be present in the parents are not in the pre-embryo that is selected. But to catch those specific mutations, you have to know in advance which mutations you are looking for. In this context, the terms "screening" and "diagnostic" describe, respectively, looking for something that could be there but you do not know will be, and something that you expect to be there based on prior information. For screens, the check is to make sure that chromosomes are not missing or duplicated to avoid conditions such as Down syndrome, which is caused by having an extra copy of chromosome 21.

The most common diagnostic evaluation is for recessive disease mutations that have been determined to be present in both parents. There are two ways that parents may know they carry mutations that could be selected against using PGT. First, they may know because they have already had a child with a recessive genetic disease. This means that each of them is a carrier for mutations that cause the disease, as, for example, in the case of Matt and Cristina Might. Each of them carries a single loss-of-function mutation in the NGLY1 gene, and Bertrand Might, the first reported patient with NGLY1 deficiency, inherited one disease-causing mutation from each of them. Without any medical intervention, every child conceived by the Mights, or by other parents like them, would have a 25 percent chance of having disease. This possibility, however, can be eliminated by testing candidate pre-embryos and ensuring that the one or ones selected for transfer carry at least one normal allele. The other way parents may know that they each carry recessive mutations to be avoided is through carrier testing, as I described earlier. Expanded carrier testing involves complete sequencing in prospective parents of around two hundred or more genes that cause recessive diseases. If mutations are observed in the same gene in the parents, the parents have the option of considering IVF and the selection of pre-embryos for transfer to the mother's uterus that are known not to carry two disease-causing mutations.

Right now, unfortunately, expanded carrier testing is often not performed, and when performed it is often later than it should have been — when a prospective mother is already pregnant. The reason it happens this way is not because it makes any sense to perform the test that late. It most certainly does not. It happens that way because this is when a lot of people find out it is possible. Their gynecologist or obstetrician explains it to them, and when a prospective mother is pregnant, she is paying more attention to such options than before she was pregnant. Many opt for the test then. But before long, most people will know to do so sooner. I hope this book might even help make that happen by raising awareness.

Determining that prospective parents carry mutations in the

same recessive disease genes before conception would increase the options available to parents, and it is not at all fanciful to presume that before long people of reproductive age will have details about their genomes loaded onto their smartphones. This would readily permit couples in everyday environments to perform the kind of genomic compatibility checks described above that take place in some religious communities. Once a large number of people of reproductive age carry their genomes on their smartphones, it is a certainty that many will perform immediate compatibility checks long before conception — even, for example, over a first drink. They will use some app that compares genomes, enter the right codes identifying their own, and read out immediately whether they carry mutations in the same genes. No doubt they will explore other things in one another's genomes as well. At a minimum, this will make for the use of more interesting vocabularies during some first dates than would otherwise be the case. At this early stage, all options are then available to the couple, from finding someone else to talk to in the bar through to conception followed by PGT to ensure that the pre-embryo selected does not have the identified pair of disease-causing mutations.

Although not yet as widely practiced, the same paradigm of pre-embryo selection could be used by parents found to carry dominant disease-causing mutations. To illustrate, it would be technologically entirely feasible for someone in the same position as Angelina Jolie to decide not to implant a pre-embryo carrying the same BRCA1 mutation that she does. Half of female pre-embryos and half of male pre-embryos would carry the mutation, and the other 50 percent of each would not. This is equally true for any other dominant mutation not on the X chromosome. As seen earlier, considering the ACMG gene list alone, we are talking about as much as 2 percent of

—

the population, or more than six million people in America alone, who carry mutations that the ACMG considers potentially life-threatening and actionable. Once these six or seven million people become aware that they carry these mutations and aware that they have the option not to transmit them to children, it seems that some would select that option. Perhaps a great many would. Although these genes in particular have been judged actionable, and so carriers do have recourse, the action that is available is not always completely effective and often not desirable.

For anyone questioning the potential long-term impact or potential societal benefit of reproductive genome design, it is important to be clear about these incontestable facts. These genetic diseases collectively are surprisingly common and constitute a tremendous global burden of morbidity and mortality. Considering only the two categories described, recessive disease affecting children and, later, dominant diseases, affected individuals run far into the millions in America alone. Greatly reducing the burden of diseases caused by such mutations seems to me a considerable contribution to human welfare. Whether efforts such as these are ultimately considered a kind of eugenics, the societal benefit still seems clear. For anyone who lives with an affected loved one or has worked with families who care for children suffering from devastating genetic diseases, being able to offer the families the option of healthy children would seem to be an unambiguous societal good. As I outlined in the introduction, however, because eugenics historically has had very different connotations, I choose to refer to such efforts with the more operational descriptor of reproductive genome design.

For anyone wondering whether large-scale reproductive genome design is restricted to the pages of science fiction, I emphasize that all I have described in the preceding few paragraphs is not only able to deliver a societal good but also definitively in technological reach. Indeed, the only pressing challenge society has in this regard, in my view, is ensuring that this benefit of modern genetics is not restricted to the global economic elite.

Other important classes of genetic causes of and contributors to disease, however, cannot be effectively addressed with today's technologies. One of the most important of these is de novo mutations. As mentioned earlier, many of the early-onset genetic diseases result from de novo mutations. In theory, the same paradigm used to address dominant and recessive mutations could be used to select pre-embryos free of de novo mutations. The problem, however, is in determining which embryos are free of such de novo mutations. Although our ability to recognize that a mutation, once identified, causes disease has improved dramatically, in part by determining the parts of the genome that do not tolerate genetic variation, sequencing technologies are not sufficient to reliably identify de novo mutations using only the DNA found in just one or a few cells as is currently available through PGT. With such a low starting amount of DNA, errors in sequencing are greater than the real rate of de novo mutations, making it impossible to reliably identify this critical class of disease-causing mutations. The only recourse as yet is to directly sequence the developing fetus, through either invasive or noninvasive sampling of fetal DNA.

The other fundamental constraint to current technologies in reproductive genome design is not only the need to detect what pre-embryos carry, but also having enough pre-embryos to choose from in order to select ones with the desired genetic characteristics, which is restricted by the number of mature eggs a woman can produce through normal reproduction or through in vitro fertilization. Human males produce a virtually unlimited supply of sperm. On

—

average, a human male produces roughly 100 million sperm cells in every teaspoon of semen. This is almost certainly more than needed to have a good chance of fertilization and very likely reflects some degree of sperm competition in the history of humans. Looking across the animal world, both the quantity of sperm and the nature of those sperm correlate with whether a male can expect his sperm to compete with that of other males in their partners' reproductive tracts. The male gorilla, relatively protected from sperm competition because the dominant male keeps other males away from the females in his troop, produces in an ejaculate roughly one one-hundredth as many sperm as does a human. Toward the other end of the spectrum, a male pig, facing more substantial sperm competition, produces sixty times as many sperm in an ejaculate as does the human male. Other species have evolved more elaborate competitive mechanisms in the competition among sperm, including the evolution of specializations such as suicide sperm, which help other sperm along, or sperm with morphological features that block the door behind them. In this context, the behavioral ecologists would view humans as having some degree of sperm competition in their evolutionary history, but relatively modest amounts compared with some species.

The production of eggs, however, is dramatically different. A human female will ovulate only about four hundred mature eggs during the course of her lifetime, one or two per month over the approximately thirty years of her reproductive life. Drugs used as a part of IVF can promote egg development to increase the quantity of the eggs available to work with, but this will amount to only about fifteen to twenty eggs per round. This fundamental constraint means that it is possible to target, for non-transmission, only a small number of variants that parents carry. This is not an important con-

—

straint for recessive diseases. As we have learned, we each carry only a few mutations in single-copy form that would cause recessive diseases if combined with another disease-causing mutation in the same gene. This is very unlikely to happen in more than a single gene in any couple. Thus only a handful of pre-embryos that are tested are almost certain to include at least one that does not have two disease-causing mutations that would result in disease. This expectation emerges directly from Mendel's rules. Recall that if both parents carry a recessive disease-causing mutation in the same gene, one-fourth of their children will have the recessive disease-causing genotype. The probability of three pre-embryos, for example, all having the disease-causing genotype is only about one in a hundred. Therefore, it is almost always possible for IVF clinics to find an embryo that does not carry two such mutations in the same gene once they know what to look for, based on what is present in the parents. We should note, however, that this simple expectation is the most optimistic case, since many pre-embryos do not develop properly and do not show a healthy morphology. Of course there is no guarantee that the genetically healthy pre-embryos will also be the ones that show the healthiest development.

These constraints on the number of pre-embryos and on the ability to accurately determine the full genome of pre-embryos today preclude IVF clinics from developing more extensive strategies in reproductive genome design. Let us return first to our thought experiment of the common-variant human. As we have discussed, all humans carry rare genetic variants that change the sequences of proteins, many of which we know are at least moderately damaging to the function of those proteins. Depending on how we define rarity, we might end up with a handful or tens of such variants in every human genome. Below I discuss the real numbers of such variants

—

in sequenced individuals (including those in my genome). For now, for simplicity, let us assume that we consider only ten such rare variants that a prospective parent wished to ensure would not be transmitted to his or her child. Assuming that they are inherited independently of one another, the probability of producing a single sperm or a single egg with no such variants would be about one in a thousand. There is virtually no chance that the few pre-embryos evaluated in the course of IVF would include any that would be free of any of those variants, considering only those from either the mother or the father, let alone both. Indeed, the sperm the father produces would include by chance some that are free of such variants. But considering all the eggs a woman might produce in her lifetime, she would not, on average, produce any that are free of rare variants, even with only ten rare variants to consider.

Sculpting predisposition to complex diseases that are heavily influenced by more common variants is even more challenging. As we have learned, many human traits and human diseases are influenced by very small contributions of thousands or tens of thousands of common variants. The exact constellation of such variants a person inherits can have profound influences or risks of diseases such as type 2 diabetes, schizophrenia, and many other such conditions. But because there are so many variants at work, the handful of pre-embryos that could be evaluated will differ only very modestly in their burden of disease due to these contributions. In fact, this conjecture has recently been directly assessed; investigators assessed the likely range represented by a set of pre-embryos in their polygenic scores for traits such as height or intelligence.[12] They found that the top-scoring pre-embryos might be two centimeters above the average in height or two points above the average in IQ. Very modest effects indeed.

These considerations might lead to the reassuring conclusion that reproductive genome design will be restricted to tackling the more straightforward and less ethically fraught application of eliminating severe Mendelian diseases. As I show in the next section, however, we are unlikely to be able to hide behind these technological limitations for long.

Reproductive Genome Design: What We Will Be Able to Do Tomorrow

As we have seen, what is possible today is restricted to selecting the preferred choice among a small number of pre-embryos. Even this modest intervention has the capacity to create dramatic changes in the genetic composition of the human population and to considerably reduce the burden of recessive genetic diseases and some forms of dominant genetic disease. But this is only a process of selection. And this selection process is further constrained by being applied only to a very small number of options among pre-embryos. Furthermore, despite dramatic advances in our ability to edit genomes at will, it is very difficult to safely edit a very small number of pre-embryos, since the rate of mistakes would likely be unacceptably high. This means, despite some recent and unfortunate forays into the genome editing of human embryos, which I will shortly describe, there is no reason to believe that the current technologies will permit us to go much further than what was outlined above. That is not to trivialize how much good could be accomplished by a more systematic application of existing technologies. There is still a long way to go in using existing technologies to help parents produce children who are as healthy as possible. Looking to the future, however, there is every reason to believe that relatively incremental tech-

—

nological advances will allow us to go much further than is possible today.

The most fundamental barrier to more extensive reproductive genome design is the limited number of pre-embryos that current IVF technologies allow. Without a larger number to work with, the simple probabilities involved (and the law of independent assortment) mean that there is no way to target more than one or two sites in the genomes of parents where you try to ensure which allele is transmitted to the pre-embryo selected for transfer. And, more generally, it is very unlikely that we could ever safely write the genomes of pre-embryos to introduce variants chosen by parents that are not already in their own genomes. But it is now clear that the barrier of a limited number of pre-embryos can eventually be overcome.

We should first emphasize that the way this barrier will be overcome is not likely to be through cloning. Cloning, of the sort that Barbra Streisand has employed to enjoy genetically similar pet dogs, originated with work by the English biologist Sir John Gurdon, who created the first cloned organism.[13] In the 1950s, Dr. Gurdon, then a graduate student, began work to extend earlier experiments that had been performed in which the nuclei of early embryo frog cells were transferred to a frog egg from which the nucleus had been eliminated. At the time, there was a consensus view that the nucleus from one embryo might allow the development of a tadpole when placed into an enucleated egg, but that this would not be true if you took a nucleus from a more mature or differentiated cell. In the parlance of the time, something restricted the activity of the DNA in the mature cell to permit the specialization of the cell, and this restriction would prohibit the nucleus of the differentiated cell from correctly specifying the development of an entire organism, in this case a tadpole. At this time, as you will recall, the structure of DNA

—

had only very recently been discovered, and there was no way to determine exactly what DNA was present in what cells. A leading hypothesis about restriction at the time was the most obvious one of all—that in the differentiated cells much of the DNA needed to construct *other* cells simply was not there. A way to test that was to see whether the experiments would work if you used not embryonic nuclei but nuclei from a differentiated cell. The beginning of what will eventually be an entirely new world of human reproduction and reproductive medicine traces to what happened when Dr. Gurdon transferred nuclei from intestinal cells in feeding tadpoles to enucleated frog eggs. The fact that the tadpoles were feeding is salient, showing that the intestinal cells were differentiated and performing their assigned differentiated roles. Remarkably, a fraction of those egg cells developed into fully functional tadpoles that could develop into normal frogs and reproduce. Clearly, whatever restricted the nucleus of the intestinal cell to behave as an intestinal cell was undone by placement in the egg. Whatever the restriction factors were, they were not loss of DNA, which of course could not be regenerated by the cell.

We now know that the environment of the egg cell in some way reprogrammed what we now refer to as the epigenetic characteristics of the transplanted nucleus to a sufficient degree to allow the DNA to regain its capacity to generate all possible frog cell types. Exactly how the cell achieves this epigenetic reprogramming magic remains largely unknown. Dr. Gurdon himself credited the histone proteins found in the egg—which as you recall, are proteins that ensconce DNA and influence which genes are and are not expressed. These experiments, when confirmed in the years that followed, were of tremendous scientific interest in demonstrating the possibility of reprogramming. And they eventually captured popular imagination

—

in the form of particular cloned individuals, most famously in the form of the first cloned mammal, Dolly the sheep. They also led to any number of explorations of the implications of cloning in the pages of science fiction novels. But with the exception of Barbra Streisand's proliferating her beloved Coton de Tulear, and some very few others who have been willing or able to spend tens of thousands of dollars to maintain genetically similar pets beyond their natural lifespans, cloning had little broader reach and none clinically. Despite the scientific importance, cloning suffers the same limitation we have discussed—and others, too. There is no ready source of human eggs, and no realistic prospect of using the technology in human reproduction.

Realization of the clinical and scientific potential of reprogramming differentiated nuclei would have to await the development of technologies that would allow a starting point without the use of an egg cell as a host. Through the many decades following Gurdon's seminal work, most scientists seem to have viewed the challenge of returning a nucleus to its beginning state as too complex to achieve without relying on whatever magic the egg environment imparted. Consequently, little systematic effort was dedicated to discovering how this might be achieved experimentally. That finally changed more than four decades after Gurdon's discoveries, when the Japanese scientist Shinya Yamanaka set out to discover how to return a differentiated cell to a state of pluripotency. Knowing that embryonic stem cells can turn into any cell type, Yamanaka studied the proteins that are critical in controlling gene expression specifically in the embryonic stem cells. His work focused on twenty-four such proteins (called transcription factors) that were known to be of particular importance in stem cells based on their representation in those cells. He was able to show that when these transcription factors were introduced into skin cells, these skin cells, remarkably, developed many of the characteristics of embryonic stem cells. By evaluating subsets of these proteins that influence gene expression, Yamanaka identified a set of only four transcription factors that can transform skin cells into cells that act like embryonic stem cells. Yamanaka had successfully reprogrammed the nuclei of the skin cells into cells that are, apparently, able to produce any kind of cell, what is now called pluripotency.

The work that Gurdon started, which has had practical culmination in the work of Yamanaka, won both of them the Nobel Prize in Physiology or Medicine in 2012 and has led to a revolution in biomedical research. The pluripotent cells generated by Yamanaka's transcription factors and related approaches are called induced pluripotent stem cells (IPSCs). In reality, Gurdon's work and its subsequent iterations, including Dolly, had already shown that reprogramming of nuclei that direct the development and behavior of differentiated cells is possible. But the reprogramming that was performed by eggs on the transferred nuclei occurred through almost entirely unknown mechanisms before Yamanaka's work. Yamanaka showed exactly how it can be done with transcription factors and, critically, made it experimentally tractable. The starting material can be as simple as a skin biopsy from a person or a blood sample. From these starting cells from any organism, including any human, it is possible to generate IPSCs that represent the genomic makeup of the donor individual. In theory, these induced stem cells can then be differentiated into any cell type (once the correct differentiation protocols are established, which will be challenging and is for many cell types very much a work in progress).

Although the potential of IPSCs in modeling human diseases and in developing truly regenerative approaches in medicine, including the possibility of replacing faulty cells and even organs, has received considerable attention, the implications for human reproduction have been less a focus to date. In fact, the implications are profound, and, I believe, will represent the next big change after IVF. It is now established that it is possible to generate functional sperm and egg cells in mice, largely or entirely in vitro. Although the differentiation protocols to allow fully functional germ cells are still evolving and there remain questions about whether the stem cells are really fully reprogrammed and whether harmful mutations are introduced in the process, it is already clear that the technologies work and can lead to viable development. There is already great interest in these technologies to help with both male and female infertility that cannot be overcome through IVF. Given the pressing clinical need in such cases, it seems certain that infertility is where these new technologies will eventually be trialed, just as was the case for IVF.

But once these technologies are worked through and applied, first to help deal with infertility, they will also open up virtually unlimited possibilities in reproductive genome design. The key reason is the most obvious one. Once human sperm and egg cells can be reliably derived from induced stem cells, there will be an unlimited supply of cells available to choose from for generating pre-embryos. This solves the problem of being able to target only a few variants for non-transmission, since it will then be possible to generate very large numbers of pre-embryos that can be screened for the desired constellation of transmitted and non-transmitted variants from the parents. But even more fundamentally, the generation of gametes from stem cells will allow the use of genome editing to write the preferred genome of one's children. In fact, genome editing can be used directly on pre-embryos, and regrettably, it has been.

But, as I describe below, there appears to be no way that this could be performed safely in any kind of general way with current technology.

Genome editing today makes use of a version of acquired immune defense that bacteria use against infection by bacteriophages, viruses that infect bacteria cells. The way the system works in bacteria is that snippets of sequence based on the infecting virus are turned into RNA, called the guide RNA. This guide RNA is then coupled to an enzyme that cleaves DNA. A common form of the enzyme that cleaves DNA is Cas9, which is a type of enzyme called an endonuclease, whose function is to slice up DNA. The guide RNAs direct the enzyme to the matching DNA sequence, which the enzyme then cuts, rendering a defense against the invading viral DNA. Through the 1990s and 2000s, this system was slowly worked out by scientists, but the real applied breakthrough came from the work of Emmanuelle Charpentier and Jennifer Doudna, who were the first to realize that the CRISPR-Cas system could be used to edit targeted genomic sequences in any kind of cell. Since their seminal realization and elaborations of the approach by them and many others, CRISPR-Cas has emerged as one of the signature tools of biomedical research, with applications throughout research labs and increasingly in gene therapy. The value is in being able to make precise changes virtually anywhere in the genome of any cell at will. This set of tools can be used to create cells and indeed mice or other organisms carrying mutations that cause disease in humans, or it can be used to correct disease-causing mutations therapeutically.

For our purposes, what the combination of CRISPR-Cas and the ability to generate gametes from stem cells means is that the core technological capacity is in place to generate stem cells, edit those stem cells to match a desired sequence, test those cells for faithful editing of the targeted sequence, and then generate gametes derived from cells with the preferred sequences. In principle, this combination means that any desired changes in the human genome become possible. It becomes possible, for example, to allow a prospective parent with only the alleles conferring Alzheimer's risk at the ApoE gene to edit the risk allele to the nonrisk allele and to generate a gamete otherwise reflecting their genomic makeup, but without the risk allele for Alzheimer's disease. And it becomes possible to undertake much more expansive sculpting of gametic genomes.

With the combination of the genome-writing tools of CRISPR-Cas and the generation of a large number of starting cells to edit through IPSCs, it becomes possible to turn the thought experiment we have used throughout this book for illustration into a reality. A parent could realistically say, "Doc, in addition to eliminating known disease variants, I would like to replace each and every rare genetic variant I carry with the common form in the human population. I don't want my child to suffer the same genetic burdens that I have myself borne." I do not mean to trivialize the difficulties of such a comprehensive sculpting of gametic genomes. We know, for example, that genome editing can create unexpected changes throughout the genome, and this is one of the reasons that the scientific community reacted with such concern and outrage when human embryos were edited in China to generate mutations in the CCR5 gene in order to confer resistance to HIV. The CCR5 gene encodes a receptor that HIV uses to enter cells, and it has long been known that there is a mutation in the human population that deletes part of the CCR5 gene, and when individuals are homozygous for this deletion, they are (largely) resistant to HIV. The scientist in China, He Jiankui, used embryos generated by IVF from a couple in which the father was HIV positive and the mother HIV negative. He then used CRISPR-Cas9 to edit the embryos and select for transfer embryos that carried the CCR5 mutation introduced by editing. He Jiankui also claimed to have checked for unwanted or so-called "off-target" effects elsewhere in the genome, though the details of this have not been made public. In fact, as we have seen, there is no way with today's technology that this could have been done reliably prior to embryo transfer. This could, today, be done reliably only post-transfer by directly evaluating the genome of the developing fetus, with the only option at that stage being selective termination if unwanted changes were introduced. Apparently, such genetic evaluation of the developing fetuses carrying the CCR5 mutations was not performed.

The outrage following these experiments was, in my view, entirely justified. First, it is generally agreed that there was not a compelling clinical need for the alterations. More fundamentally, as we have already described, currently the genetic analysis of a handful of cells from an embryo is not sufficiently accurate to allow us to determine whether unwanted mutations have been introduced elsewhere in the genome. The potential benefit to the children soon to be born in no way justified the risks that were being taken. And it is exactly this trade-off that must be considered when we move new technologies into clinical applications with real people and real consequences. It is difficult for me to see the experiment as anything other than scientific opportunism involving unacceptable risks. And I frankly blame all those who directly or indirectly enabled the work, in addition to He Jiankui. This scientific and clinical objection is compounded in this case by an apparent violation of well-accepted ethical standards of requiring both the truly informed consent of the research subjects and the appropriate institutional review of an entirely new and highly risky experiment *on human beings*. In this

—

case, it appears that neither appropriate informed consent nor appropriate institutional review took place. This is not a good beginning to the editing of human embryos. It is also a critical warning to the community about the things that might happen as the technologies of reproductive genome design evolve.

My own view, and that of virtually all working scientists, is that the editing of embryos is currently far too risky to be practiced. We simply do not yet have the ability to be sure of what we will have done to the embryos, and no way to be certain of the outcome. When we use CRISPR-Cas9 editing in a research setting, we always edit a large number of cells and identify the ones that have the desired changes. And critically, in best practice we also confirm the absence of unwanted changes somewhere else in the genome that are not being targeted. This is simply not possible to do in the editing of pre-embryos with current technologies. And probably it will not soon be possible using only a handful of embryos as a starting point, even for a single site edited as performed by He Jiankui, let alone more extensive genome design as described above.

But once we consider the generation of gametic cells from stem cells, suddenly it becomes possible to make changes, and to be sure that those changes, and only those changes, are introduced. This is because it will become possible to select—among the virtually unlimited number of stem cells that can be generated, edited, and sequenced—exactly those stem cells that have the desired changes, and only the desired changes. Although substantial development and the subsequent fine-tuning are still to come, we can already see the technologies in plain view that will permit this. I suspect it will be many years before these methods are sufficiently reliable to be used for human reproduction, but the basic technological ingredients are not only possible but here, or at least nearly here. It is therefore time to decide how they are to be used. This book is, above all, meant to make this prospect a real enough possibility to encourage serious thought about reproductive genome design so that when the time comes, in however many years it takes, we are as ready as we can be.

Reproductive Genome Design: What Should We Do?

Based on what we know today, it would be difficult to predict with any accuracy when extensive reproductive genome design will become a reality. But what we can say is that it is coming. At some point. And if the rate of recent developments in cellular reprogramming and genome editing is any guide, it seems to me the time will come within the next ten to thirty years — an interval that can easily pass well before we are ready to make wise use of such abilities.

Perhaps the best-known analogy in science, and consequently the most overused (second only to the moon landing), is the Manhattan Project. In this case, the analogy seems particularly fitting. The Allied powers recognized that advances in physics might permit the development of a bomb of previously unimaginable destructive capability, so they set out to develop their own before the Germans did. What is less well known is the extreme simplicity of the initial call to action, penned by the physicist Leo Szilard, signed by Albert Einstein, and submitted to President Franklin D. Roosevelt. The physicists noted simply that "it may become possible to set up a nuclear chain reaction in a large mass of uranium, by which vast amounts of power and large quantities of new radium-like elements would be generated." They did not have all the details worked out, and they noted that the amount of uranium required would

—

likely make such a bomb not transportable by air. Of course, six years later the *Enola Gay,* a Boeing B-29 bomber named for the mother of its pilot, carried the first atomic bomb used in battle and dropped it on the Japanese city of Hiroshima. Despite the simplicity of the letter, the physicists nevertheless outlined what was possible, eventually leading to the Manhattan Project. And with the general theoretical basis already clear, the Manhattan Project overcame remaining technological hurdles and in a few years developed the first atomic weapons.

We find ourselves in an analogous situation, but hopefully without the tragic consequences. The broad theoretical outlines of how comprehensive reproductive genome design can be carried out are now known and can be stated as simply as Szilard and Einstein did for the prospects for an atomic weapon. We now know that it will be possible to generate human gametes starting from adult cells that are reprogrammed into induced pluripotent stem cells. And we know that those stem cells can be edited through CRISPR-Cas9 editing to stem cells carrying whatever desired genetic changes the parents may select before differentiation into gametes. In this case, the analogy is entirely apt. We know how this can be done. And in knowing how it can be done, we know that it will be done.

To be clear, that this reality will come is, to me, where the Manhattan Project analogy ends. It is not my opinion that we should, as did the scientists involved in the Manhattan Project, work to overcome the remaining technological hurdles as quickly as we possibly can. Quite the contrary. The American effort to develop a bomb was initiated in fear that the Germans would get there first. I see no comparable reason to push to overcome, as quickly as possible, the remaining technical hurdles to achieving comprehensive genome design. Indeed, as I hope this book has persuaded you, we do not yet know how to wisely make use of such technological abilities. Instead, my point is that in terms of the broad picture, we know this is coming, and it is time to think hard about how we want to use the technology. Indeed, my hope is that if the technologies mature quickly, we will be sufficiently prepared to know, at least, what we should not do.

So let us continue the analogy now and fast-forward to August 6, 1945, when the *Enola Gay* dropped on Hiroshima the results of the Manhattan Project. This day for reproductive genome design will come, and we can begin even now to outline what could be done, knowing what we already know. What might be done with the new technological abilities?

We have already seen how future generations can be protected from mutations responsible for devastating diseases, both those that are recessive and those that are dominant. The advent of more comprehensive genome design will allow us to do this more efficiently than is possible today. But, as we have seen, application of existing technologies, if fully deployed, could protect future generations from many of these disease-causing mutations. One critical category that cannot be adequately addressed currently is that of de novo mutations that cause severe disease. And in a new reproductive model, using genomically screened stem cells selected to be free of severe mutations before being differentiated into gametes, this will become possible. Ensuring that children are not born with such mutations, or any others causing devastating genetic disease, is clearly a positive contribution to human well-being. This particular application, it should be noted, can also be achieved currently through pre-natal diagnostics, which in principle can be used to test fetuses for de novo mutations, as already is done to some degree. And increasingly this is and will continue to be done through noninvasive approaches that

—

determine the genetic makeup of a fetus by sequencing cell-free DNA circulating in the mother's blood. Currently this approach is feasible only for a portion of disease-causing mutations, but with reductions in sequencing costs it will increasingly allow complete determination of the genetic makeup of the developing fetus. But this approach allows only the fraught binary choice of continuing a pregnancy or selectively terminating it.

Reproductive genome design, on the other hand, combined with the use of edited stem cells differentiated into gametes, will allow us to go well beyond protecting future generations from mutations that cause devastating disease. If parents had this option now, what might they do?

———

We can now return to our thought experiment and ask, Will parents decide that they wish their children not to have any of the small genomic blemishes that they themselves carry? This now becomes straightforward by, for example, using CRISPR-Cas or similar editing technologies to edit each and every site in the stem cells generated from a parent to swap the rare genetic variants they carry for more common ones. To make this concrete, let us consider the type and number of variants we are talking about. At the Columbia Institute for Genomic Medicine (IGM), which I direct, we have sequenced the genomes of about 120,000 people over the years. This sequencing has included patients with a range of different common and rare diseases, and sometimes the parents of affected children. In order to assess the number and type of variants that might be "targeted" in a real version of our thought experiment, consider 2,664 of those individuals that we have sequenced at the IGM that do not have any known disease. How many variants of the sort we are talking about do they carry?

Let us first consider the very rarest variants. These are variants that are seen in an individual's genome but never seen in any other individual. Given the number of individuals whose genomes have been sequenced today, complete absence means that a variant is very rare indeed—not seen in any (other) of our own 120,000 individuals that we have sequenced and also not seen in any of the individuals in even larger external databases known as the Exome Aggregation Consortium (ExAC) and the Genome Aggregation Database (gnomAD). All analyses that geneticists can run suggest that these rarest of variants are also the ones most likely to damage protein function, most likely to be negatively selected in the human population, and most likely to cause disease. For example, as some highly selective illustrations of these claims, we know that the rarer a protein-coding variant is, the more likely it is to represent a non-conservative amino acid change. This means, for example, that an amino acid that is hydrophobic (because it is supposed to sit in the middle of a cell membrane) is swapped for a hydrophilic one (that does not tolerate sitting in a cell membrane at all).

In these 2,664 individuals, let us consider only the less than 2 percent of the human genome that directly encodes protein. In that small portion of the human genome, let us further consider only variants that either change the amino acid sequence or result in no protein being made at all. Considering this very extreme subset of rare functional genetic variation, we find that the individuals carry an average of approximately nineteen such variants.[14] And while the average is nineteen per individual, we see a range from zero to forty-eight. If we relax this frequency to focus on variants seen only

—

once in every 10,000 individuals, still very rare variants indeed, the average number jumps to fifty-eight per person, now with a range from eighteen to ninety-four.

You might reasonably wonder: What do these variants do in these individuals, and why are they there anyway? The answer to the first question should not surprise you after reading the earlier chapters, particularly chapter 3. The truth is that we simply do not know what these individuals would have been like if they had not had those variants. We do know, as I have emphasized previously, that such low-frequency variants are very strongly enriched for deleterious effects on protein function and enriched for variants that selection acts against.

The question of why they are there is more easily addressed. They are surely present largely because of what geneticists refer to as a mutation-selection balance. The variants present in the population represent a balance between those introduced by mutation and those removed by selection. Although some of these variants are no doubt neutral, many are under at least modest negative selection, and such variants remain present in low frequency in the human population because they are being introduced each and every generation, in each and every person, by new mutation. And as long as the variant is compatible with life, it can sometimes take a few generations for even a fairly strongly deleterious variant to be removed by natural selection.

As noted earlier, I confirmed the uncertainty about the consequences of removing such variants by asking geneticists to predict what we call the phenotype of such a common-variant human. Among the roughly ten professional geneticists that I asked, very few answers matched, even in very broad outline. One of my own students, now a very famous professor, suggested that the resulting offspring would be dead; a couple suggested that they would be "mostly average." Consistent with that answer, James Watson, who of course figured prominently in this history, immediately offered up the thought that a common-variant human would have an IQ of 100 — that is, bang-on average, reflecting his or her carrying a full complement of common variants.

My own guess is that the male version of common-variant man might look something like Chris Hemsworth, the actor who plays Thor, at least in height, physique, and apparent overall fitness (but without reference to skin color). I come to this view on the supposition that rare variants, even when not causal of recognized genetic diseases, are often somewhat debilitating. Some may make us a little shorter than we would otherwise be, or in some cases rather too tall. Some may adversely affect our blood pressure, eyesight, stamina, lung function, kidney function, or intelligence. Remove those rare variants, and you would have someone who is fairly tall but not a giant, with good lungs, kidneys, heart, and overall health. And, to put it more succinctly, Chris Hemsworth has a role in the Marvel Universe instead of, say, the geneticist writing this book, not because he carries the god of thunder gene but because he got fairly lucky in the constellation of slightly damaging rare variants he inherited. Indeed, consistent with this view, it is also my guess that people who live very long, healthy lives are not generally protected by a "longevity gene," but rather got lucky in having relatively few rare damaging variants that lead to relatively earlier common or rare diseases.

I promised at the outset of the book to hazard my own guess about what a common-variant human would look like, and my guess is Thor. But I have also promised to label my opinions, as opposed to established scientific consensus, and this very much qualifies.

—

Many geneticists would ascribe much more importance not to the lottery of rare damaging variants but to the lottery of more common variants. And the varied answers I received to the thought experiment emphasize precisely that uncertainty. Indeed, when pressed, several of those offering guesses adjusted their answers. The prominent geneticist who claimed that the common-variant human would not survive admitted that since the replacement variants would all be the common form in the population, there is no reason to apply the death sentence, but did not hazard further phenotyping beyond commuting the death sentence.

Watson himself, on my suggesting that the constellation of rare variants we all carry is likely to be somewhat damaging to various complex traits, adjusted his answer to an IQ of 120. Whether for the same reason or not, this is largely in accord with my own guess regarding IQ, that it would be high but not off-the-charts high (which is not to make any comment on Thor's IQ). The reason I lean toward a high but not truly exceptional IQ is that I suspect that selection for increased intelligence has been very strong in most of human history, and this is the reason we are a great deal smarter than any other species on earth. For this reason, any gene variants that increase IQ without adverse consequences have probably already swept through human populations. The variants still present in the human population that strongly affect IQ, I suspect, often reduce it. This is not to say that there are no common variants with modest effects on IQ, and indeed many have been identified. But those with strong positive effects and no adverse consequences are already the dominant or only form in our genomes and precisely the reason humans are very smart.

Under this hypothesis, the truly unusually gifted may be influenced by variants that increase intelligence but at the price of sometimes causing adverse effects that are selected against, at least in some people in which the variants are found. And indeed this perspective is one of my two principal reasons for rejecting Watson's conjecture about differences in intelligence among humans with different geographic ancestries being genetically mediated. Selection has, I think quite obviously, been for higher intelligence in all human groups. There is no reason whatsoever for one group to have more "high-IQ" variants than another. The second reason I already outlined in the opening pages. Watson likes to emphasize that individuals of different ancestries perform differently on standardized tests, but it is extremely well established that test performance depends heavily on environment. That remains by far the most parsimonious explanation for such differences.

I should be clear that my guesses about common-variant humans are really just that. I have no way of proving those guesses and, while they are my best guesses, even I do not have much confidence in the outcome. Certainly not enough confidence to entertain eliminating such variants that I carry in my own children, were I to have had the chance. But to make clear that we all do carry such variants, I will reveal that my own genome has six variants that change the protein sequences of six different genes that are never seen in any other individual, either in our own data or in any public database. What these variants, and other slightly less rare variants, may have to do with my precise makeup and general good health at fifty-six (poor eyesight notwithstanding), I have no idea. And whether the fact that I have slightly fewer such variants than average has anything to do with who I am I also have no way of knowing, despite spending most of my waking time on the study of human genetics.

As the varied answers of geneticists make clear, we really do not know what would happen if we generated a common-variant human being. My point is not to convince you that my guess is right. Instead, my fundamental point is that we do not know what would happen if we replaced all the rare variants with more common alleles, but there are reasons to believe that many traits we care about would be affected in some way. And I find it hard to believe that as this becomes possible some people will not choose to perform the experiment. Perhaps even those who find that they have rather more of these questionable rare variants than the average person.

The question of what should be done with heritable human genome editing (HHGE) was recently addressed by a committee established jointly by the National Academy of Medicine and the National Academy of Sciences in the United States and the Royal Society in the United Kingdom. The report is technically unimpeachable, and the advice offered, in my view, comprehensively sensible. While there are many detailed recommendations, they can perhaps be reasonably distilled to arguing that HHGE should be undertaken only when the safety profile of editing is well established, only to prevent serious monogenic diseases, and only when selection among pre-embryos is unable to prevent disease.

Although I am generally sympathetic to these recommendations, I do not believe it likely that they will be followed once the technology advances sufficiently to permit parents to choose to make more aggressive alterations. To be sure, the path will be established by interventions along the indicated lines, beginning with devastating diseases that cannot be otherwise addressed. But there will be interest in going further as soon as it is technically possible, and the recent experience of HHGE in China should make clear to all that what is possible will be performed somewhere. And beyond this

—

consideration, the report avoids some of the most challenging questions, such as, What should be considered a disease? The reality is much messier than what is implied by simply restricting attention to "single-gene disease." The most fundamental question we might ask is, What is a disease?

The most authoritative treatise on psychiatric disorders is the *Diagnostic and Statistical Manual of Mental Disorders,* now in its fifth edition. This consensus view about mental disorders is published by the American Psychiatric Association, and used by clinicians and drug regulators alike to determine what diseases are appropriate for treatment and for drug development. But what is considered a disease does not have fixed standards and is obviously influenced by prevailing sentiment and prejudice. Remarkably, from our current vantage point, homosexuality was included as a psychiatric disease in the *DSM* until 1973, and then, after it was removed as a disease, it continued to exist until 1987 in a murky category of sexual orientation disturbance. This tragic and clearly prejudiced history should make us skeptical of any attempt to hide behind the idea of restricting our attention exclusively to recognized diseases.

Another problem is the aim to focus only on single-gene disorders. We can easily recognize some qualifying diseases, many of which we have discussed. It is equally clear, however, that many humans are influenced by genetic variants that operate in a vast gray zone between what geneticists like to call single-gene and complex diseases. How are we to ensure that parents target only the former? And should geneticists be making that decision for them? I find these questions essentially unanswerable currently, and feel that the only certainty is that as the technology becomes available, some people, somewhere, will use it. And we would then learn the answers only by effectively experimenting with human beings. It seems to

—

me today that the only defense against widespread unintended consequences is to be as informed as possible. And right now, we are a long way from being informed.

At the time of this writing we have no way of knowing the consequences of such changes, but I believe we can know that some parents, if offered the choice, will decide to engineer their children in some of the ways discussed here, in the hope of having children that are smarter, stronger, taller, more attractive, longer lived, and even richer than they themselves are.

And we can further imagine some parents using the bewildering set of polygenic risk scores that have appeared in the literature in an attempt to mold the constellation of common variants that an embryo carries in order to minimize the risk that these common variants collectively carry of schizophrenia, epilepsy, cardiovascular disease, and any number of other diseases and traits, including height and even sexual orientation. And of course, using only what we know now, parents could order up cosmetic changes in traits including eye color, hair color, skin pigmentation, and others. To make this as concrete as possible, just imagine what parents would say if asked what they would like the probability to be that their child will develop schizophrenia based on his or her genetic makeup. In almost all cases the answer would be, "As low as possible." And if given the choice, might very tall parents ask to have their children's polygenic complement of variants edited so as to result in children of more normal height, and might some shorter parents ask for the reverse? This seems a certainty.

And what would be the harm in efforts to so design the genomes of our children? There are two answers to this question, which I consider the defining questions of the future of human reproduction. The first answer relates to what we think is both beneficial and ethical to do. But this is a question best addressed once we better know the outcomes we might be able to safely achieve. The second, and pressing, question is about the potential harm we might do as we experiment with the genomes of our children. Take the example of our thought experiment, where, as I have outlined, geneticists clearly have no idea what would happen.

And what of the other major category that could be immediately addressed with our new Manhattan Project capabilities in reproductive genome design? Given the very diverse set of polygenic risk scores now available, what would happen if parents were to specify preferences from a large menu of options regarding polygenic risks for different diseases and traits? Quite reasonably, they might select the minimization of cardiovascular disease, autism, epilepsy, schizophrenia, and type 2 diabetes – and, for good measure, throw in some tall-gene variants for parents of below average height? What would be the harm? In fact, the polygenic risk scores are reliable for exactly what they are. They do indeed confer the indicated risks for the indicated conditions (with the caveat outlined below). But to think about the consequences, you have to think about exactly how these scores have been derived.

Take schizophrenia as an example. Recent scores have been derived by comparing the common variants found in about forty thousand patients with schizophrenia compared with about a hundred thousand population controls. From this comparison, variants associated with increased risk of schizophrenia are identified. And, collected over a large number of such variants, the differences in risk can be profound. For example, if you take individuals who are in the top tenth percentile in terms of their constellation of risk variants and compare them with those in the bottom tenth percentile, the difference in the probability of having schizophrenia is almost

—

five-fold. But you can probably already see the problem. The study is focused on schizophrenia and schizophrenia alone. What else is different about those who are in the upper tenth decile and those in the lower tenth decile? In fact, we currently have virtually no idea. The studies are simply not performed in such a way as to answer such questions (with some small-scale exceptions). And there are other buried and insidious problems with the use of such scores in reproductive genome design. More than ten years ago, Anna Need and I summarized the geographic ancestry of individuals that were included in large-scale genomic studies.[15] We found that the research subjects included in such studies were overwhelmingly of European ancestry. In fact, at that time, there were nine individuals with European ancestry included for every individual with non-European ancestry. Unfortunately, this extreme and simply unacceptable disparity has continued, despite efforts at remediation. Recently, Alicia Martin and Mark Daly and their colleagues have shown that polygenic risk scores do not carry over reliably across different ancestry groups.[16] Parents, therefore, might discover that the risk scores they are using will not work as advertised in the context of the ancestry of their own particular genomes.

So where does this leave us? In fact, not in a comfortable place at all, and being in that place is the reason for this book. Genetics, as we have known it, is likely coming to an end, but what will replace it? As it looks now, our technological ability to design the genomes of our children is set to far exceed our knowledge of the consequences of that redesign. And concerns about the potential consequences in fact extend well beyond the uncertainty about the nature of human genetic variation that I have tried to outline here. We do not yet know whether the changes that could be undertaken might yield generally more fit humans, but we also cannot know what unintended consequences might emerge. For example, we have little way of knowing how the bonds between parents and their children might be affected by wholesale genome engineering. Do we really know how much the universal bonds that parents and children feel are influenced by a sense of shared identity and how much they might be affected by scrubbing the genomes of children of some of the quirks of parental individuality? Beyond the technical challenges we have focused on, the question of how genomics might represent very real risks to our sense of biological belonging in families represents new and unpredictable risks as we enter into a new, designed form of human genetics. There are old affinities that structure relationships and might have unpredictable consequences as they are changed.

Also, given the recent history of eugenics, it is difficult to be sanguine about how the widespread availability of genetic information might be used. When most people have genomic information on their smartphones, what happens when a phone is hacked and the information disseminated? Will carriers of noteworthy mutations be ridiculed on social media, even if laws seek to prevent discrimination? As even my brief remarks in the earlier chapters should have made clear, even very influential scientists are far from immune to the biases that so often inflame prejudiced perspectives. As we consider how very broad dissemination of genetic information might or might not be, we would do well to recall that in 1966, easily within living memory, Linus Pauling suggested that individuals with the mutation responsible for sickle cell disease should have a warning "tattooed on the forehead of the carriers, so that young men and women would at once be warned not to fall in love with each other." He also suggested that carriers who marry non-carriers should be restricted in the number of children they can have.[17]

That these suggestions could be expressed only two decades

after the end of World War II, complete with harrowing echoes of the tattoos that Jewish and other genetic "undesirables" still carried with them from concentration camps, should be a clarion call to us all that genetic information is not safe, even in the hands of the most distinguished. Pauling was, after all, not only a Nobel Prize-winning chemist but also the winner of the Nobel Peace Prize. I believe that modern genetics has helped us to see through some of these cruel and uninformed eugenic prescriptions. As you learned earlier, nearly all of us carry single disease-causing mutations in genes that cause recessive diseases, and I do not believe we will again hear arguments that such individuals should be restricted in the number of children they should be allowed to naturally conceive. But before we applaud ourselves too loudly for the progress we have made, we should remember that unfounded claims about racial differences in genetically determined intelligence are still promulgated to this day, including by another Nobel laureate, as discussed earlier. And it is very hard for me to believe that Pauling's prescriptions for dealing with sickle cell were unaffected by the fact that those variants are concentrated in individuals of African ancestry.

We might hope that society will simply wait until we, collectively, know what we are doing. And by knowing what we are doing, I refer to all of the attendant challenges and concerns. These challenges include the technical challenges I have emphasized throughout the book, as well as the challenge of likely misuse of genetic information, which the history of genetics over the past century should make us aware is all too likely. Even though the specific misrepresentations of science in the service of bias will surely change, we can safely assume the phenomenon will continue. And I refer also to questions of how relationships among family members might be affected by the use of reproductive genome design.

—

But I consider this a forlorn hope, that society will wait until it is ready, technically and otherwise, to assume a direct role in guiding the genomic future of our species. Once we pass our Hiroshima moment in reproductive genome design, parents will try some of the options I have already outlined, and no doubt myriad others. As the example in China makes clear, once it becomes possible, experiments will be performed, with people. To me, above all, this sets up a race between our technological abilities in reproductive genome design and our knowledge of the consequences. Since the technological ability is on the way, there are many things we should do to prepare ourselves. Many of those things are outside the scope of this book and my own expertise. But one that is within my scope is developing a much richer and more nuanced understanding of the consequences of differences in human genetic variation. Indeed, this is our Manhattan Project. Designing the genomes of children is on the way. But first, using all the focused energy and ingenuity we can muster, we need to work out exactly what we ought to do and ought not to do. Ultimately, much of the data required for this already exists in the variation present on earth, if systematically studied. If this book ultimately has a clear conclusion, it is that we must make a much more systematic effort to understand the consequences of human genetic variation. While this is only one of the challenges that we will face in reproductive genome design, it is a key one and one that we know how to address.

How might we increase our understanding of the role of genetic variation in determining the differences among people? What can and should be done is in fact clear. It can be summarized quite succinctly by incorporating complete genomic information into the health care provided to all people. This might sound a rather outlandish suggestion. But even from a purely economic perspective, it is an eminently prudent investment. The recent coronavirus pandemic has shown, among other things, the profound vulnerability of global economic activity to global health. As I write, scientists throughout the world (including me, when not typing away) are racing to sequence the genomes of patients who suffer severe disease when infected with SARS-CoV-2 in order to determine whether there are genetic differences among people that influence their susceptibility.

The reason for this work is well known to geneticists and well illustrated by the HIV/AIDS pandemic. HIV enters a cell by binding to its receptor, encoded by the CCR5 gene. Fifteen percent of northern Europeans, however, carry a deletion in this gene, and individuals who carry two copies of this deletion are almost completely protected against HIV. A drug that inhibits this protein was later shown to be effective therapeutically in HIV-positive patients. Might there be something like this genetic difference in human genomes that would provide us guidance in the treatment of COVID-19? Now a full year into this crisis, with 163 million global cases reported and 3.37 million deaths, we still have no idea. Today scientists scramble to collect DNA samples from patients, then to find the resources to generate sequence data for those patients, and then to slowly build up collections to analyze and compare with those of other institutions.

Fortunately, it already appears that vaccines are beginning to control the pandemic in early 2021. But next time we may be faced with a pathogen that is not as amenable to vaccines as SARS-CoV-2 seems to be, with its relatively low mutability. And indeed, even with the relatively low mutability of SARS-CoV-2 compared with some other viruses, already variants that may partially evade existing vaccines are a concern. Imagine, however, that all Americans had already been sequenced, before the pandemic started, and that their sequence data were stored in a protected repository that they or their health-care proxies could grant access to. Patients coming into the hospital to be treated for COVID-19 could simply say yes or no to providing access to their genetic data for research, using an opt-in system like the organ donor system connected to state driver's licenses throughout the United States. Had this system been in place before the COVID-19 pandemic, we would have known at the beginning whether there is a CCR5-like discovery lurking in our genome for SARS-CoV-2.

For years, scientists have debated whether such comprehensive genomic analysis is worth the cost. From the perspective of delivering health care to meet current needs, the answer has been unclear and the debate fraught. But from the perspective of making our economies more resilient in the context of infectious diseases alone, the answer seems clear to me. Considering only the American context, it might cost on the order of thirty billion dollars to generate sequence data for the most important parts of the genomes of all Americans. This is a fraction of the cost the US government has already invested only in relief measures for the pandemic. The world admirably applies tremendous resources to health care, with the United States dedicating nearly 20 percent of its twenty-trillion-dollar economy annually to health care. Generating sequence data on Americans would clearly position the country to be more resilient in the face of future global health crises. These considerations illustrate, I believe, that generating comprehensive sequence data for a large proportion of the world population is not only feasible but economically prudent. And it will be just such data that we need to ensure that as we enter a new age of reproductive genome design, we will have as much information as possible to guide the choices of parents. Only

—

genetic data on such a scale will really teach us how genetic variation influences who and what we are. We need to collect these data because reproductive genome design is on the way. And for those making economic arguments, I believe the COVID-19 pandemic alone makes clear that it is economically sensible to do so.

———

If there is any single message in this book, it is that we must truly understand how the genetic differences among the billions of people on earth influence their lives and health before we select the genomes of our children. That is our challenge, and it is our obligation. And the time to start is now.

NOTES

CHAPTER 1. THE FUTURE OF REPRODUCTION

1. Lehmann-Haupt, "Want a Perfect Baby?"

2. Ghorayshi, "This Company Is Trying to Make More Perfect Babies."

3. Wade, "A Dissenting Voice as the Genome Is Sifted to Fight Disease."

4. Smith, "Exaggerations and Errors in the Promotion of Genetic Ancestry Testing"; BBC Radio 4, "Uncovering British 'Deep Ancestry.'"

CHAPTER 2. LEARNING TO READ THE HUMAN GENOME

1. Fisher, "Has Mendel's Work Been Rediscovered?" 115–37.

2. Avery, MacLeod, and McCarty, "Studies on the Chemical Nature of the Substance Inducing Transformation of Pneumococcal Types," 137–58.

3. Watson and Crick, "Molecular Structure of Nucleic Acids," 737–38.

4. The Cambridge scientists did not, in fact, get permission from Franklin to interpret her data on Photo 51 (DNA under X-ray crystallography), which, to them, quickly suggested the double-helix structure. Maurice Wilkins, though he famously did not get along with Franklin, worked with her at King's College and showed Watson Photo 51. Watson and Crick also were able to review a progress report of work in the Franklin lab that had been provided to Max Perutz. Whether this report should have been treated as confidential has been a matter of debate.

—

Franklin herself died before the Nobel Prize was awarded, making it unclear whether she would have been included, since the award is not given posthumously.

◀

CHAPTER 3. THE NATURE OF
HUMAN GENETIC VARIATION

1. Wakap et al., "Estimating Cumulative Point Prevalence of Rare Diseases," 165–73.

2. Goldstein, "Heterozygote Advantage," 1195–98.

3. The mathematical tools used in population genetics remained largely attributed to these three pioneers until the development of coalescent theory in the 1980s. Classical population genetics can be thought of as describing the evolution of populations through time in the normal "forward" direction. Thus, a classical population genetic problem might focus on characterizing how long it takes, on average, for a population of a given size to lose one of the genetic forms in the population due to genetic drift. To conceptualize such a problem, classical population genetics would imagine the alleles from one generation to make up the gene forms of the next generation. Coalescent theory approached questions in population genetics from essentially the opposite perspective. Instead of conceptualizing populations moving forward in time, coalescent theory focuses on only the allelic forms present in the current generation and develops models for calculating expected ancestral relationships among those forms based on the characteristics of the population. For many problems, coalescent theory is a far more efficient framework for gleaning results and now provides a critical tool kit for deriving population genetic results and for making inferences about population history from measurable genetic data.

CHAPTER 4. DNA AND HUMAN DISEASE

1. Botstein et al., "Construction of a Genetic Linkage Map in Man," 314–31.

2. In fact, no genetic variants have yet been identified that influence weight through an effect on energy efficiency. Instead, all known disorders of weight relate to satiation signaling.

3. Goldstein, "Common Genetic Variation and Human Traits," 1696–98.

4. Boyle, Li, and Pritchard, "An Expanded View of Complex Traits," 1177–86.

5. National Research Council, *Mapping and Sequencing the Human Genome*.

—

6. National Research Council, *Toward Precision Medicine.*

7. Olson, "A Behind-the-Scenes Story," 3–10.

CHAPTER 5. WRITING THE GENOMES OF OUR CHILDREN

1. With thanks to Joe Hostyk in the Goldstein lab for scraping results from Online Mendelian Inheritance in Man (OMIM), at www.omim.org, March 2020. National Institutes of Health, "Clinical Genomic Database."

2. Baird et al., "Genetic Disorders," 677–93.

3. With thanks to Joe Hostyk in the Goldstein lab for scraping results from OMIM.

4. I say generally because, as noted briefly in chapter 4, de novo mutations usually occur in the gamete leading to the embryo and are not otherwise discernible in any parental cells. Sometimes, however, the responsible mutations occur in a broader set of cells in one of the parents. This can happen when the mutation is what we call post-zygotic in the parent. It may be present in a small fraction of parental cells, including some of the cells leading to gametes. In this case, we refer to the parent as mosaic. It has been estimated that for de novo mutations causing diseases in children, in as much as 10 percent of the cases it is possible to find the responsible mutations present in the parent but at a relatively low representation. This means that the parent does not have the disease because most cells do not carry the mutation, but it means that the risk of recurrence for future offspring is high.

5. American College of Medical Genetics and Genomics, "The Use of ACMG Secondary Findings," 1467–68.

6. Jolie, "My Medical Choice."

7. Rego et al., "High-Frequency Actionable Pathogenic Exome Variants in an Average-Risk Cohort."

8. Liu et al., "Apolipoprotein E," 106–18.

9. Abondio et al., "Genetic Variability," 222.

10. Karavani et al., "Screening Human Embryos for Polygenic Traits," 1424–35.

11. Here I am ignoring the use of artificial insemination first known to have been performed in the 1700s.

12. Karavani et al., "Screening Human Embryos for Polygenic Traits," 1424–35.

13. Stevens, "Barbra Streisand Cloned Her Dog."

14. With thanks to Gundula Povysil for help with the number of variants seen in genomes sequenced in the Institute for Genomic Medicine (IGM), including my own.

15. Need and Goldstein, "Next Generation Disparities in Human Genomics," 489–94.

16. Martin et al., "Current Polygenic Risk," 584–91.

17. Linus Pauling, August 15, 1966, quoted in Gormley, "Eugenics for Alleviating Human Suffering"; Everett Mendelsohn, "The Eugenic Temptation: When Ethics Lag Behind Technology," *Harvard Magazine,* March 1, 2000, harvardmagazine.com/2000/03/the-eugenic-temptation.html.

BIBLIOGRAPHY

Abondio, Paolo, Marco Sazzini, Paolo Garagnani, et al. "The Genetic Variability of APOE [Apolipoprotein E] in Different Human Populations and Its Implications for Longevity." *Genes* 10, no. 3 (March 2019): 222. https://doi.org/10.3390/genes10030222.

American College of Medical Genetics and Genomics (ACMG) Board of Directors. "The Use of ACMG Secondary Findings Recommendations for General Population Screening: A Policy Statement of the American College of Medical Genetics and Genomics (ACMG)." *Genetics in Medicine* 21, no. 7 (July 2019): 1467–68. https://doi.org/10.1038/s41436-019-0502-5.

American Psychiatric Association. *Diagnostic and Statistical Manual of Mental Disorders (DSM-5), Fifth Edition.* Washington, DC: American Psychiatric Association, 2013.

Avery, Oswald T., Colin M. MacLeod, and Maclyn McCarty. "Studies on the Chemical Nature of the Substance Inducing Transformation of Pneumococcal Types: Induction of Transformation by a Desoxyribonucleic Acid Fraction Isolated from Pneumococcus Type III." *Journal of Experimental Medicine* 79, no. 2 (February 1944): 137–58. https://doi.org/10.1084/jem.79.2.137.

Baird, Patricia A., Theodore W. Anderson, Howard B. Newcombe, and Brian R. Lowry. "Genetic Disorders in Children and Young Adults: A Population Study." *American Journal of Human Genetics* 42, no. 5 (May 1988): 677–93.

Bateson, William. *The Methods and Scope of Genetics: An Inaugural Lecture Delivered 23 October 1908.* Cambridge: Cambridge University Press, 1908.

BBC Radio 4. "Uncovering British 'Deep Ancestry.'" Last upated July 9, 2012. http://news.bbc.co.uk/today/hi/today/newsid_9736000/9736128.stm.

BIBLIOGRAPHY

Bell, Callum J., Darrell L. Dinwiddie, Neil A. Miller, et al. "Carrier Testing for Severe Childhood Recessive Diseases by Next-Generation Sequencing." *Science Translational Medicine* 3, no. 65 (January 2011): 65ra4. https://doi.org/10.1126 /scitranslmed.3001756.

Botstein, David, Raymond L. White, Mark Skolnick, and Ronald W. Davis. "Construction of a Genetic Linkage Map in Man Using Restriction Fragment Length Polymorphisms," *American Journal of Human Genetics* 32, no. 3 (May 1980): 314–31.

Boyle, Evan A., Yang I. Li, and Jonathan K. Pritchard. "An Expanded View of Complex Traits: From Polygenic to Omnigenic," *Cell* 169, no. 7 (June 2017): 1177–86. http://doi.org/10.1016/j.cell.2017.05.038.

Chiurazzi, Pietro, and Filomena Pirozzi. "Advances in Understanding – Genetic Basis of Intellectual Disability," *F1000Research* 5, F1000 Faculty Rev (April 2016): 599. https://doi.org/10.12688/f1000research.7134.1.

Cochran, Gregory, and Henry Harpending. *The 10,000 Year Explosion.* New York: Basic, 2009.

Crow, James F. "Mid-Century Controversies in Population Genetics." *Annual Review of Genetics* 42 (2008): 1–16. https://doi.org/10.1146/annurev.genet.42.110807 .091612.

Fisher, R. A. "Has Mendel's Work Been Rediscovered?" *Annals of Science* 1, no. 2 (April 1936): 115–37. https://doi.org/10.1080/00033793600200111.

Fukuyama, Francis. *The End of History and the Last Man.* New York: Free Press, 1992.

Ghorayshi, Azeen. "This Company Is Trying to Make More Perfect Babies." *BuzzFeed News,* July 12, 2015. www.buzzfeednews.com/article/azeenghorayshi/more -perfect-babies.

Goldstein, David B. "Common Genetic Variation and Human Traits." *New England Journal of Medicine* 360, no. 17 (April 2009): 1696–98. http://doi.org/10.1056 /NEJMp0806284.

——. "Heterozygote Advantage and Evolution of a Dominant Diploid Phase," *Genetics* 132, no. 4 (December 1992): 1195–98.

Gormley, Melinda. "Eugenics for Alleviating Human Suffering." *It's in the Blood! A Documentary History of Linus Pauling, Hemoglobin, and Sickle Cell Anemia,* page 35. Special Collections and Archives Research Center, Oregon State University Libraries, 2015. http://scarc.library.oregonstate.edu/coll/pauling/blood/narrative/page 35.html.

GWAS (Genome-Wide Association Studies). "GWAS Catalog: The National Human Genome Research Institute (NHGRI) Catalog of Human Genome-Wide Association Studies." European Bioinformatics Institute (EMBL-EBI). Accessed November 27, 2020. www.ebi.ac.uk/gwas/.

Hayden, Erika C. "Should You Edit Your Children's Genes? In the Fierce Debate About

BIBLIOGRAPHY

CRISPR Gene Editing, It's Time to Give Patients a Voice." *Nature* 530, no. 7591 (February 2016): 402–5. http://doi.org/10.1038/530402a.

Jolie, Angelina. "My Medical Choice." *New York Times*, May 14, 2013. www.nytimes .com/2013/05/14/opinion/my-medical-choice.html.

Karavani, Ehud, Or Zuk, Danny Zeevi, Nir Barzilai, et al., "Screening Human Embryos for Polygenic Traits Has Limited Utility." *Cell* 179, no. 6 (November 2019): 1424–35. https://doi.org/ 10.1016/j.cell.2019.10.033.

Kerr, Gabriel. "Family Says Precision Medicine Provided New Hope." *ABC News*, August 2, 2016. https://abcnews.go.com/Health/family-precision-medicine-provided -hope-sick-daughter/story?id=41048386.

Kohler, Robert E. *Lords of the Fly: Drosophila Genetics and the Experimental Life.* Chicago: University of Chicago Press, 1994.

Lehmann-Haupt, Rachel. "Want a Perfect Baby? Counsyl Says: Just Spit." *CBS News*, February 23, 2010. www.cbsnews.com/news/want-a-perfect-baby-counsyl-says -just-spit/.

Liu, Chia-Chen, Takahisa Kanekiyo, Huaxi Xu, and Guojun Bu. "Apolipoprotein E and Alzheimer Disease: Risk, Mechanisms and Therapy." *Nature Reviews Neurology* 9, no. 2 (February 2013): 106–18.

Martin, Alicia R., Masahiro Kanai, Yoichiro Kamatani, Yukinori Okada, Benjamin M. Neale, and Mark J. Daly. "Clinical Use of Current Polygenic Risk Scores May Ex-acerbate Health Disparities." *Nature Genetics* 51, no. 4 (April 2019): 584–91.

Meghan, the Duchess of Sussex. "The Losses We Share." *New York Times*, November 25, 2020. www.nytimes.com/2020/11/25/opinion/meghan-markle-miscarriage .html.

Mendel, Gregor J. "Versuche über Pflanzenhybriden." *Verhandlungen des naturforschen-den Vereines in Brünn, Bd. IV für das Jahr, 1865. Abhandlungen* (1866): 3–47. https://doi.org/10.5962/bhl.title.61004.

Mnookin, Seth. "One of a Kind: What Do You Do If Your Child Has a Condition That Is New To Science?" *New Yorker,* July 14, 2014. www.newyorker.com/magazine /2014/07/21/one-of-a-kind-2.

National Academy of Medicine, National Academy of Sciences, and the Royal Society. *Heritable Human Genome Editing.* Washington, DC: The National Academies Press, 2020.

National Institutes of Health (NIH). "Clinical Genomic Database." National Human Genome Research Institute. Last updated September 22, 2020. https://research .nhgri.nih.gov/CGD/.

National Research Council. *Mapping and Sequencing the Human Genome.* Washington, DC: National Academies Press, 1988.

———. *Toward Precision Medicine: Building a Knowledge Network for Biomedical Research and a New Taxonomy of Disease.* Washington, DC: National Academies Press, 2011.

Need, Anna C., and David B. Goldstein. "Next Generation Disparities in Human Genomics: Concerns and Remedies." *Trends in Genetics* 25, no. 11 (November 2009): 489–94. https://doi.org/10.1016/j.tig.2009.09.012.

Olson, Maynard V. "A Behind-the-Scenes Story of Precision Medicine." *Genomics Proteomics Bioinformatics* 15, no. 1 (February 2017): 3–10. https://doi.org/10.1016/j.gpb.2017.01.002.

Rego, Shannon, Orit Dagan-Rosenfeld, Wenyu Zhou, M. Reza Sailani, Patricia Limcaoco, Elizabeth Colbert, Monika Avina, et al. "High-Frequency Actionable Pathogenic Exome Variants in an Average-Risk Cohort." *Cold Spring Harbor Molecular Case Studies* 4, no. 6 (December 2018): a003178. http://doi.org/10.1101/mcs.a003178.

Simpson, George G. "Tempo and Mode in Evolution." *Transactions of the New York Academy of Sciences* 8 (December 1945): 45–60. https://doi.org/10.1111/j.2164-0947.1945.tb00215.x.

Smith, David. "Exaggerations and Errors in the Promotion of Genetic Ancestry Testing." *Genomes Unzipped.* Accessed February 2, 2021, https://genomesunzipped.org/exaggerations-and-errors-in-the-promotion-of-genetic-ancestry-testing/.

Stevens, Matt. "Barbra Streisand Cloned Her Dog. For $50,000, You Can Clone Yours." *New York Times,* February 28, 2018. www.nytimes.com/2018/02/28/science/barbra-streisand-clone-dogs.html.

Wade, Nicholas. "A Dissenting Voice as the Genome Is Sifted to Fight Disease." www.nytimes.com/2008/09/16/science/16prof.html.

———. "Genes Show Limited Value in Predicting Disease." *New York Times,* April 15, 2009. www.nytimes.com/2009/04/16/health/research/16gene.html.

Wakap, Stéphanie N., Deborah M. Lambert, Annie Orly, et al. "Estimating Cumulative Point Prevalence of Rare Diseases: Analysis of the Orphanet Database," *European Journal of Human Genetics* 28, no. 2 (February 2020): 165–73. https://doi.org/10.1038/s41431-019-0508-0.

Watson, James D., and Francis H. Crick. "Molecular Structure of Nucleic Acids: A Structure for Deoxyribose Nucleic Acid." *Nature* 171, no. 4356 (April 1953): 737–38. https://doi.org/10.1038/171737a0.

INDEX

INDEX

INDEX

Huxley, Julian, 78
hybrid vigor, 68, 71

immune system, 119
inborn errors of metabolism, 83
inbreeding, 112
incidental findings, 115–16
individuality, genomic, 108–9, 120–21
infertility, 123, 138. *See also* in vitro fertilization (IVF)
inheritance: blending, 36–37, 38, 45; X-linked, 111–12
intellectual disability, 114, 115
intelligence, 64, 132, 149, 150–51, 158
interventions. *See* treatment
intolerance scoring, 17, 101–3
in vitro fertilization (IVF), 122–27, 130, 131, 132, 134, 138
IPSCs (pluripotent stem cells), 137, 140, 144
Ireland, 120
isolation, 65
IVF (in vitro fertilization), 122–27, 130, 131, 132, 134, 138

Jenkin, Fleeming, 45
Jews, 121–22, 158
Jolie, Angelina, 116, 127
Judson, Horace Freeland, 46

kidney disease, 100, 115
Kornberg, Arthur, 54

labels, clear, 39
Lander, Eric, 88
law of dominance and uniformity, 37
law of independent assortment, 41, 48
law of segregation, 39
linkage, 84–85, 86, 89, 99
linkage disequilibrium, 89–90

malaria, 69, 73
Manhattan Project, 143–44, 145
maps, genetic: common variant map, 90; first, 48–49; of genetic markers, 84; of human genome, 81–82
Marfan's syndrome, 114
Martin, Alicia, 156
Mendel, Gregor, 23, 35–45, 58, 113; arguments against, 41–43; confirmation of laws, 47; exceptions to laws, 44–45; laws of, 37, 39, 41, 44, 48, 111, 131; rediscovery of, 45, 46–47
mental disorders, 153
Meselson, Matt, 53
meta-analysis of results, 42
Miescher, Friedrich, 51
Might, Bertrand, 15–18, 99, 107, 113, 126
miscarriages, 101–3
Modern Synthesis, 59, 77–78
Moffat, Alistair, 13–14
molecular biology, 51
molecular clock, 74, 75, 76
Morgan, Thomas Hunt, 47
mosaic mutation, 98, 165n4
Muller, Hermann, 72–73, 74, 79
mutations: actionable, 115–16, 117, 128; de novo, 7, 86, 98–99, 129, 145–46, 165n4; dominant disease-causing, 113–20, 127; fixing before birth, 107; frameshift, 56; harmful, 32; major-effect, 14–15; Mendelian diseases and, 73; natural selection and, 59; post-zygotic, 165n4; recessive, 44, 70, 111–13; on same chromosome, 48. *See also* variants
mutation-selection balance, 148

National Academy of Medicine, 152
National Academy of Sciences, 152

INDEX

INDEX